Since its founding, NASA has been dedicated to the advancement of aeronautics and space science. The NASA Scientific and Technical Information (STI) Program Office plays a key part in helping NASA maintain this important role.

The NASA STI Program Office is operated by Langley Research Center, the lead center for NASA's scientific and technical information. The NASA STI Program Office provides access to the NASA STI Database, the largest collection of aeronautical and space science STI in the world. The Program Office is also NASA's institutional mechanism for disseminating the results of its research and development activities. These results are published by NASA in the NASA STI Report Series, which includes the following report types:

- TECHNICAL PUBLICATION. Reports of completed research or a major significant phase of research that present the results of NASA programs and include extensive data or theoretical analysis. Includes compilations of significant scientific and technical data and information deemed to be of continuing reference value. NASA counterpart of peer-reviewed formal professional papers, but having less stringent limitations on manuscript length and extent of graphic presentations.

- TECHNICAL MEMORANDUM. Scientific and technical findings that are preliminary or of specialized interest, e.g., quick release reports, working papers, and bibliographies that contain minimal annotation. Does not contain extensive analysis.

- CONTRACTOR REPORT. Scientific and technical findings by NASA-sponsored contractors and grantees.

- CONFERENCE PUBLICATION. Collected papers from scientific and technical conferences, symposia, seminars, or other meetings sponsored or co-sponsored by NASA.

- SPECIAL PUBLICATION. Scientific, technical, or historical information from NASA programs, projects, and missions, often concerned with subjects having substantial public interest.

- TECHNICAL TRANSLATION. English-language translations of foreign scientific and technical material pertinent to NASA's mission.

Specialized services that complement the STI Program Office's diverse offerings include creating custom thesauri, building customized databases, organizing and publishing research results ... even providing videos.

For more information about the NASA STI Program Office, see the following:

- Access the NASA STI Program Home Page at *http://www.sti.nasa.gov*

- E-mail your question via the Internet to help@sti.nasa.gov

- Fax your question to the NASA STI Help Desk at (301) 621-0134

- Phone the NASA STI Help Desk at (301) 621-0390

- Write to:
  NASA STI Help Desk
  NASA Center for AeroSpace Information
  7121 Standard Drive
  Hanover, MD 21076-1320

NASA/TM-2006-214300

Aerocapture Systems Analysis for a Neptune Mission

Mary Kae Lockwood, Karl T. Edquist, Brett R. Starr, Brian R. Hollis, and Glenn A. Hrinda
NASA Langley Research Center, Hampton, Virginia

Robert W. Bailey, Jeffery L. Hall, Thomas R. Spilker, Muriel A. Noca, N. O'Kongo, and Robert J. Haw
Jet Propulsion Laboratory, Pasadena, California

Carl G. Justus and Aleta L. Duvall
Computer Sciences Corporation, Huntsville, Alabama

Vernon W. Keller
NASA Marshall Space Flight Center, Marshall Space Flight Center, Alabama

James P. Masciarelli, David A. Hoffman, Jeremy R. Rea, Carlos H. Westhelle, Claude A. Graves
NASA Johnson Space Center, Houston, Texas

Naruhisa Takashima
AMA, Inc., Hampton, Virginia

Kenneth Sutton
National Institute of Aerospace, Hampton, Virginia

Joseph Olejniczak, Y. K. Chen, Michael J. Wright, and Bernard Laub
NASA Ames Research Center, Moffett Field, California

Dinesh Prabhu
ELORET Corporation, Sunnyvale, California

R. Eric Dyke
Swales Aerospace, Hampton, Virginia

Ramadas K. Prabhu
Lockheed Martin Engineering and Sciences Company, Hampton, Virginia

National Aeronautics and
Space Administration

Langley Research Center
Hampton, Virginia 23681-2199

April 2006

TABLE OF CONTENTS

NEPTUNE AEROCAPTURE SYSTEMS ANALYSIS

Mary Kae Lockwood

NASA Langley Research Center, Hampton, Virginia, 23681-2199

A Neptune Aerocapture Systems Analysis is completed to determine the feasibility, benefit and risk of an aeroshell aerocapture system for Neptune and to identify technology gaps and technology performance goals. The high fidelity systems analysis is completed by a five center NASA team and includes the following disciplines and analyses: science; mission design; aeroshell configuration screening and definition; interplanetary navigation analyses; atmosphere modeling; computational fluid dynamics for aerodynamic performance and database definition; initial stability analyses; guidance development; atmospheric flight simulation; computational fluid dynamics and radiation analyses for aeroheating environment definition; thermal protection system design, concepts and sizing; mass properties; structures; spacecraft design and packaging; and mass sensitivities.

Results show that aerocapture can deliver 1.4 times more mass to Neptune orbit than an all-propulsive system for the same launch vehicle. In addition aerocapture results in a 3-4 year reduction in trip time compared to all-propulsive systems. Aerocapture is feasible and performance is adequate for the Neptune aerocapture mission. Monte Carlo simulation results show 100% successful capture for all cases including conservative assumptions on atmosphere and navigation. Enabling technologies for this mission include TPS manufacturing; and aerothermodynamic methods and validation for determining coupled 3-D convection, radiation and ablation aeroheating rates and loads, and the effects on surface recession.

SYMBOLS/NOMENCLATURE

A	=	Area (m^2)
α_{trim}	=	Trim Angle of Attack
CA	=	Axial Force Coefficient
CBE	=	Current Best Estimate
CD	=	Coefficient of Drag
CFD	=	Computational Fluid Dynamics
CG, cg	=	Center of Gravity
CL	=	Coefficient of Lift

CN	=	Normal Force Coefficient
D	=	Drag
GA	=	Gravity Assist
L	=	Lift
L/D	=	Lift-to-Drag ratio
M/CDA	=	Ballistic Coefficient (kg/m^2)
SEP	=	Solar Electric Propulsion
TPS	=	Thermal Protection System

INTRODUCTION

AEROCAPTURE significantly increases the mass that can be delivered in orbit at a destination with an atmosphere compared to an all-propulsive vehicle at the same destination with the same launch vehicle. Aerocapture utilizes aerodynamic forces on a vehicle during a single pass through a destinations atmosphere to capture into orbit about that destination, instead of a large propulsive delta V maneuver. An aerocapture flight profile schematic showing the primary aerocapture event sequence is shown in Fig. 1.[1]

Aerocapture at Neptune is characterized by high entry velocities (28-30 km/sec inertial) into a H_2 He atmosphere, and capture into a high energy science orbit enabling Titan flybys. Table 1 provides a comparison of the Neptune aerocapture reference mission, described in this paper, to a representative Mars aerocapture mission, and a Titan aerocapture reference mission[1]. The high entry velocities at Neptune compared to Titan and Mars result in significantly more severe environments at Neptune, including both aeroheating and g's. The high energy science orbit for Neptune

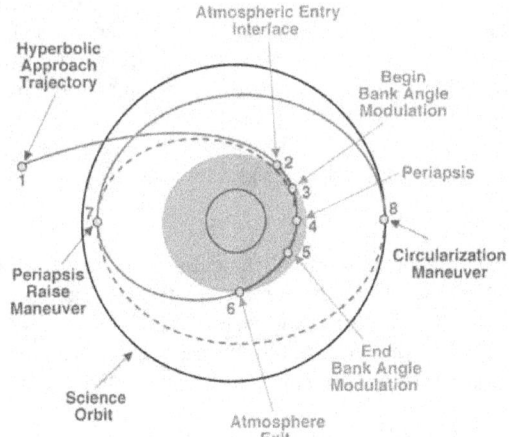

Figure 1. Aerocapture trajectory schematic.

compared to the reference Titan and Mars missions, requires a significantly greater vehicle lift to drag ratio to provide adequate corridor width at Neptune.

Table 1 Neptune aerocapture parameters compared to those at Titan and Mars.

	Neptune	Titan	Mars
Entry Velocity (km/sec)	29	6.5	5.7
Nom. Entry Flight Path Angle (deg)	-12.818	-36	-14.2
Apoapsis/Science Orbit (km)	3986 x 430,000*	1700	1400
Atmosphere Composition (% volume)	80% H2, 19% He, 1% CH4	95% N2, 5% CH4 (max)	95.3% CO2, 2.7 %N2
Atmos Scale Height at Aerocapture Alt (km)	49	40	10.5
Atmospheric Interface Altitude (km)	1000 (above 1 bar)	1000	125
Aerocapture Altitude (km)	100-300 (above 1 bar)	200-400	40
Aerocapture Exit/Escape Velocity	.97	.69	.76
L/D	.8	.25	.25
M/CDA (kg/m^2)	895	90	148
Theoretical Corridor (deg)	2.27	3.5	~1.4
Time from Atmos Entry to Atmos Exit (min)	10	42	10
Convective Stag Point Heat Rate (W/cm^2)	8000	46 (.91 m nose rad)	30 (1.9 m nose radius)
Radiative Stag Point Heat Rate (W/cm^2)	4000-8000	93-280	Negligible
Max g's During Aerocapture (Earth g's)	22	3.5	2.5-3

* For set up of Triton flyby resonance at 488,000 or 393,000 km apoapsis

APPROACH

A multi-center aerocapture systems analysis team, including NASA engineers and scientists from Ames Research Center (ARC), the Jet Propulsion Laboratory (JPL), Johnson Space Center (JSC), Langley Research Center (LaRC), and Marshall Space Flight Center (MSFC), led by Langley Research Center, was kicked off in October 2002 and completed in October 2003. The effort was funded through the Code S In Space program.

The mission objectives and initial spacecraft design for the reference concepts are based on JPL's TeamX study[2] of the Neptune Orbiter with probes mission. From this starting point, further science definition and initial analyses are completed to provide understanding of the vehicle requirements and selection of the reference concept and mission. Higher fidelity analyses are completed on the reference concept including mission design; aeroshell configuration screening and definition; interplanetary navigation analyses for determination of approach navigation delivery dispersions; atmosphere modeling; computational fluid dynamics (CFD) for aerodynamic performance and database definition; initial stability analyses; guidance development; atmospheric flight simulation; CFD and radiation analyses for aeroheating environments; TPS design, concepts and sizing; mass properties; aeroshell and spacecraft structural design and sizing; spacecraft design and packaging; and mass sensitivities.

SCIENCE

The Neptune mission includes a Neptune orbiter and two probes. The orbiter science mission includes two years in Neptune orbit. The science orbit is selected to enable Triton flybys. The Neptune Orbiter science instruments were selected to be representative, and include visible imager, IR imaging spectrometer, UV imaging spectrometer, thermal-IR imaging spectrometer, ion and neutral mass spectrometer, magnetometer, charged-particle detector, plasma wave spectrometer, microwave radiometer, USO (radio occultations) and two identical probes.[3]

MISSION DESIGN AND REFERENCE CONCEPT SELECTION

Many alternate mission designs are considered,[4] including launch on Delta IVH and Atlas 551; gravity assists utilizing various combinations of Venus, Earth and Jupiter; SEP at various power levels or chemical stages; and aerocapture versus chemical insertion. Launch dates after 2015 are considered, to provide time for technology development. An SEP, aerocapture system is baselined for the reference architecture.

The reference mission selected is a compromise between trip time, net delivered mass, inertial entry velocity, theoretical corridor width and aeroheating. Fig. 2 and 3 illustrate the net delivered mass and entry velocity vs. flight time for a range of SEP/aerocapture concepts. In general, as flight time decreases the net delivered mass decreases and entry velocity increases. The Delta IVH VJGA trajectories are selected for the reference mission concept based on delivered mass capability. Trip times less than 10 years are eliminated due to the rapid decrease in delivered mass capability and rapid increase in entry velocity (and corresponding aeroheating) with shorter trip times.

To further select a trip time, entry velocity, and required vehicle L/D, an initial trade in available corridor width as a function of vehicle L/D and entry velocity is completed. Fig. 4 shows the theoretical corridor width vs. entry velocity and L/D for aerocapture into a 350,000 km apoapsis orbit at Neptune. The results in Fig. 4 illustrate several points. Theoretical corridor increases with both L/D and entry velocity. An L/D = 0.8 vehicle at 28 km/sec provides approximately the same theoretical corridor width as an L/D = 0.6 vehicle at slightly greater than 30 km/sec. In addition, to achieve reasonable theoretical corridor widths for aerocapture into the high energy elliptic orbit requires vehicles with significantly greater L/D than the high heritage blunt body configurations. (Ex., The theoretical corridor width is only approximately 0.8 degrees for an L/D = 0.25 for a 70° sphere cone.)

To provide an initial estimated theoretical corridor required for comparison to the available corridor width, a quick corridor margin analysis is completed for combinations of L/D = 0.6, 0.8 and 28 and 30 km/sec entry velocity. The theoretical corridor must be adequate to accommodate dispersions, uncertainties, and variability in approach navigated states at atmospheric interface, aerodynamics, atmosphere, and guidance robustness. Initial estimates for navigated errors show ±.5° 3σ errors in entry flight path angle at 28 km/sec and ±.6° 3σ errors in entry flight path angle at 30 km/sec. Aerodynamic uncertainties can be conservatively estimated to

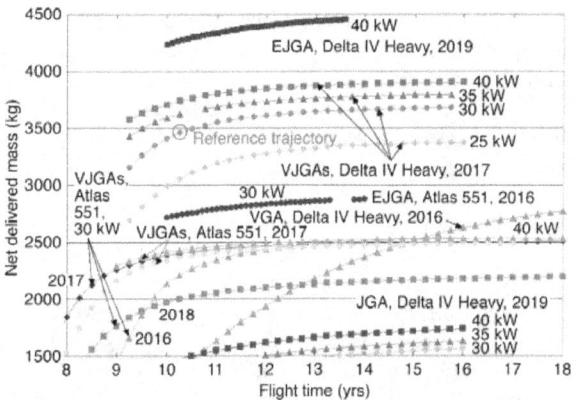

Figure 2. Delivered mass vs. trip time for a range of SEP/Aerocapture mission concepts considered. [4]

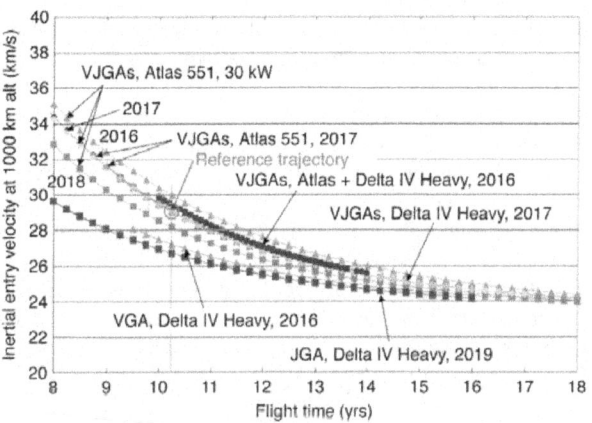

Figure 3. Entry velocity vs. trip time for a range of SEP/Aerocapture mission concepts considered. [4]

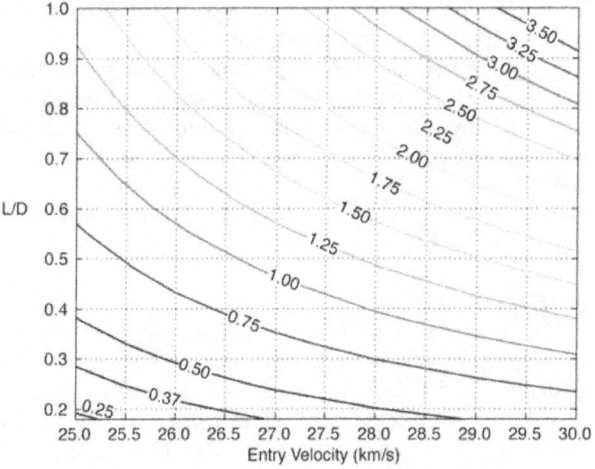

Figure 4. Theoretical corridor width available as a function of vehicle L/D and entry velocity.

3

result in ±.2 L/D, guidance is estimated to capture 95% of theoretical corridor, and corridor loss due to total mean variability of the atmosphere is estimated using lift up and lift down trajectories at the global extremes of mean density in the initial NeptuneGRAM atmosphere model. Note that this approach does not account for high frequency variability in atmospheric density and the corresponding impact on vehicle performance and margin.

Results of this estimate are shown in Fig. 5 for the 4 cases considered, L/D = 0.6 and 28 km/sec, L/D = 0.6 and 30 km/sec, L/D = 0.8 and 28 km/sec, L/D = 0.8 and 30 km/sec. The L/D = 0.6 and 28 km/sec case show the estimated corridor loss due to approach navigated errors, aerodynamics uncertainty, and total mean atmosphere variability and uncertainty. If these losses are RSS'd a required theoretical corridor width can be estimated for use in comparison to the total available theoretical corridor width.

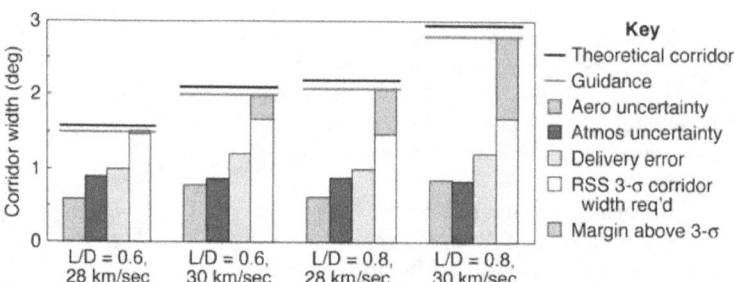

Figure 5. Estimated theoretical corridor width required compared to theoretical corridor width available.

For this case, the required and available theoretical corridor widths are approximately the same, well within the accuracy of the estimate, and again, high frequency atmosphere perturbations are not included. Because of the early phase of design, the objective for selection of the reference concept is to select a concept with margin greater than the RSS required corridor width. As a result, an L/D = 0.8 vehicle with a 29 km/sec entry velocity is selected for the reference concept, with an L/D = 0.6 vehicle kept as an option.

The reference concept is therefore described as follows. The mission launches February 17, 2017, on a Delta IV H. The launch vehicle fairing is 5 m in diameter with a 4.572 m static payload diameter. The total launch capability is 5964 kg, with a launch C3 of 18.44 km2/sec². The SEP system is a 30 kW EOL, 6 engine SEP system, that operates to 3 AU. A Venus, Jupiter gravity assist is utilized. The total trip time is 10.25 years, with Neptune arrival in 2027. Two probes are released at E-4 months (1 week apart). The probes enter at E-4 hours and E-2 hours. Aerocapture inertial entry velocity at Neptune is 29 km/sec, atmospheric interface is 1000 km above 1 bar. The orbit is 157° retrograde, 430,000 km by 3986km. The science mission includes two years in Neptune orbit for a total 12.25 year Neptune Orbiter mission.

AEROSHELL CONFIGURATION

Aeroshell configuration screening is completed to develop shapes with L/D = 0.8, and L/D = 0.6 as an option, while maximizing volumetric efficiency and minimizing M/CDA. Several aeroshell shape classes are considered including ellipsleds, flattened ellipsleds, high fineness sphere cones, biconics and bent biconics, as shown in Fig. 6.

The vehicle mass and volume are fixed, the shapes within each vehicle configuration class are varied parametrically. Newtonian aerodynamics, varified with CFD, is utilized to screen the configurations for L/D and M/CDA over a range of angle of attack. Packaging efficiency is also screened through determination of vehicle volumetric efficiency. Based on these analyses, the flattened ellipsled is selected for the reference aeroshell configuration. Details of the configuration screening and vehicle selection are contained in Ref. 5.

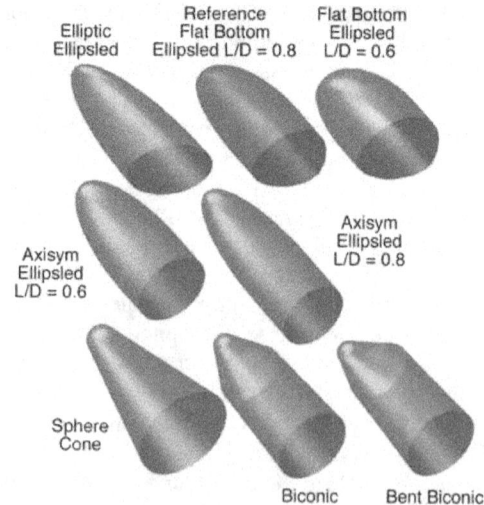

DESIGN CYCLES

Two design cycles are completed for the Neptune Orbiter. The original objectives were to package 3 probes within the aeroshell of the orbiter. As a result the design cycle one vehicle is 5.5 m in length with M/CDA estimated at 273 kg/m³, and an aeroheating design trajectory developed at 400 kg/m2 to provide mass growth margin. With this design, the system mass margin on the Delta IVH SEP VJGA was estimated to be less than the desired 35%. In addition, further analysis shows that the science objectives can be met with the probes carried externally and re-

Figure 6. Configuration classes considered.

leased prior to aerocapture. Two external probes are therefore included in the design cycle two concept. The aeroshell is photographically scaled from a 5.5 m length vehicle to a 2.88 m length, shown in Fig. 7. The entry allocation is 2200 kg, resulting in a ballistic coefficient of 895 kg/m². This M/CDA is used for both the performance analyses and the design trajectories for the aeroheating and structure design.

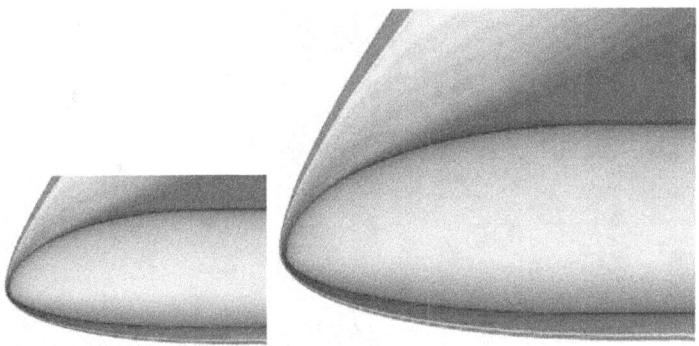

Figure 7. Comparison of Design Cycle 1 and Design Cycle 2 configuration scale.

NAVIGATION

Navigation analyses are completed at 28 and 30 km/sec with the Mars Reconnaissance Orbiter camera and an Entry – 3 day data cut-off. In addition, results are completed with an MRO camera with two times the pictures and with an advanced MRO camera. As a comparison results are also completed for an Entry – 2 day data cut-off. Results are shown in Fig. 8. See Ref. 6 for detailed discussion on the navigation analysis and results.

Monte Carlo simulations are completed for the reference vehicle using ±.51° 3σ entry flight path angle dispersions. This is approximately equivalent to an MRO camera with 2x pictures or to an advanced MRO camera, each with a more conservative Entry -3 day data cut-off. A change to Entry -2 day cut-off significantly reduces the delivery entry flight path angle dispersions as shown in Fig. 8.

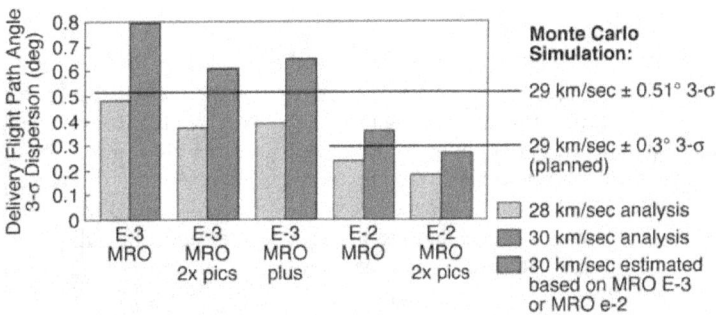

Figure 8. Delivery flight path angle dispersions at atmospheric interface for various navigation scenarios.

ATMOSPHERE MODELING

A NeptuneGRAM[7] atmosphere model is developed based on Voyager and other data. Variability includes all measurement uncertainty, residual uncertainty due to turbulence and waves, and the expected variability due to latitude, altitude, seasonal and time of day variations. The atmosphere composition is 80% H2, ~19% He, ~1% CH4. Fig. 9 illustrates the total mean density variability as a function of altitude. Note that the range of aerocapture altitudes is between approximately 100 and 300 km. The parameter Fminmax is utilized to define the range of density profiles. The mean density profile is represented by Fminmax = 0, the minimum density profile is represented by Fminmax = −1, and the maximum density profile is represented by Fminmax = +1.

Fig. 10 illustrates the latitudinal variation of density for the particular arrival season of the reference concept. This variation of Fminmax with latitude is represented by

$$Fminmax = 0.44*\cos(4.0*latitude) + fbias$$

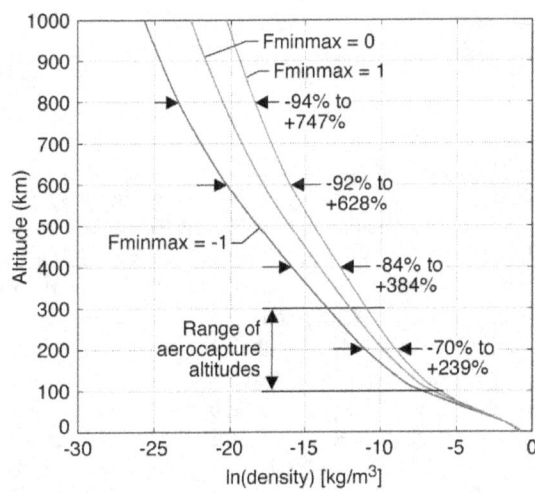

Figure 9. NeptuneGRAM mean density profile variability.[8]

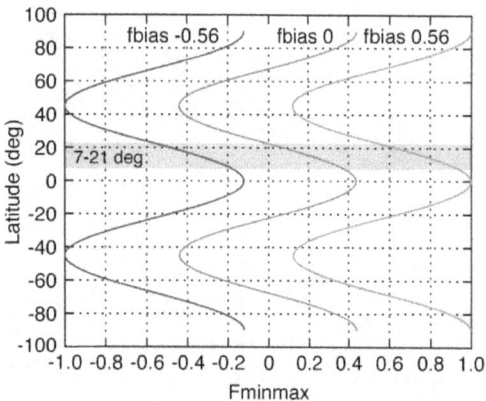

Figure 10. Effect of latitude on NeptuneGRAM mean density profile for Neptune Orbiter arrival date. [8]

where –0.56 > fbias < 0.56

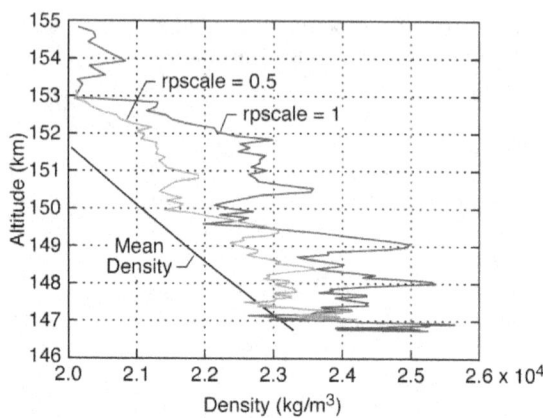

Figure 11. High frequency density perturbations in NeptuneGRAM. [8]

Therefore, for a typical Neptune aerocapture trajectory flying through the atmosphere between 7-21° latitude, $-0.6 \geq$ Fminmax ≤ 0.93, compared to a range of $-1.0 \geq$ Fminmax ≤ 1.0 for a global variation.

Fig. 11 illustrates a sample high frequency density perturbation compared to the mean density. The mean density corresponds to a given Fminmax value in Fig. 9. rpscale controls the high frequency variability of the atmosphere, with rpscale = 1 representing the greatest expected variability for Neptune. The rpscale = 1 results in Fig. 9, show a sample of how the high frequency content can alter the mean variability. Note that the high frequency content can act to increase or decrease the mean density with altitude, in addition to adding the high frequency content. Rpscale = 0.5 represents a potential decreased high frequency content for the Neptune atmosphere.

The reference concept performance, as shown below, is based on the latitudinal variation of Fminmax and rpscale = 1. Aeroheating and structure design trajectories are based on the full range of Fminmax and rpscale = 1.

AERODYNAMICS

The aerodynamic database is developed from viscous LAURA CFD analysis of the reference configuration. [5] The vehicle trims at 40° angle of attack with an axial cg location relative to the vehicle length of 0.51 aft of the nose, and a vertical cg relative to the vehicle length of .0166 below the vehicle waterline. For the trimmed vehicle L/D = 0.806, CD = 1.405, CL = 1.133. Initial stability analysis shows that the flat-bottom ellipsled is longitudinally and laterally stable. [5]

The aerodynamic uncertainties are based on the JSC ellipsled analysis for Mars, consistent with the X-33 aerodynamic database uncertainty model in Ref. 9. As shown in Fig. 12, CA: ±0.048 and CN: ±0.12, each using base area as the reference. The trim angle of attack uncertainty is assumed to be ±4°, defined in this initial analysis to be double that for a typical blunt body, such as a 70° sphere cone. Cg uncertainties are ±0.5% for axial cg relative to the vehicle length and ±0.125% for radial cg relative to the vehicle length. Based on stacked aerodynamic uncertainties, the L/D uncertainty is +26.4% and –22%. Based on an RSS of the aerodynamic uncertainties, the L/D uncertainty is +13.5% and –14.3%. The Monte Carlo variability for 2001 cases is between the RSS and stacked uncertainties. [9]

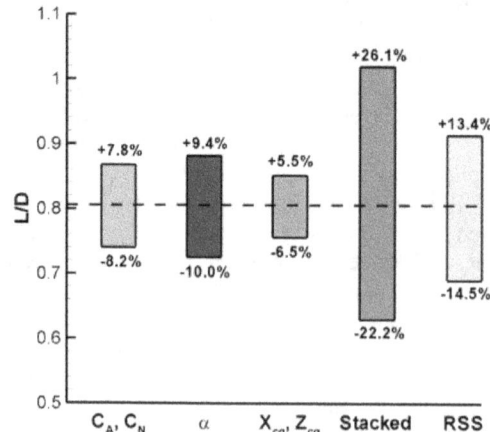

Figure 12. Aerodynamic uncertainties. [5]

Note that the effects of large TPS recession and resultant shape change on the vehicle aerodynamics and cg location have not been quantified. This analysis was outside the study scope.

GUIDANCE

HYPAS guidance (ref. 10) was chosen for the Titan aerocapture systems analysis. HYPAS utilizes vehicle lift and bank angle control through the atmosphere to target the desired exit orbit apoapsis and inclination. It is an analytically derived algorithm based on deceleration due to drag and altitude rate error feedback. This analytic, non-iterative, on-the-fly approach leads to efficient code (~320 source lines in Fortran), minimal storage requirements, and fast and consistent execution times.

HYPAS consists of two phases: 1) Capture Phase: Establishes pseudo-equilibrium glide conditions; 2) Exit Phase: Exit conditions are predicted analytically assuming a constant altitude rate followed by constant acceleration. The lift vector is adjusted to null the error between predicted and target apoapsis, and bank reversals are used to keep inclination errors within the desired limits. Results show excellent performance and an ability to capture ~93% of the theoretical corridor.

PERFORMANCE/SIMULATION

The reference concept performance is simulated in a Monte Carlo simulation[8] and includes each of the uncertainties and dispersions as described above. Fig. 13a-d show the reference concept Monte Carlo results. The reference concept is an L/D = 0.8 vehicle, with $M/CDA = 895$ kg/m^2. The target orbit, to enable Triton flybys, is retrograde with an apoapsis of 430,000 km, and a periapsis of 3986 km. Uncertainties included in the Monte Carlo include navigation, with ±0.51° 3σ entry flight path angle at atmospheric interface, atmosphere variability as a function of latitude and high frequency perturbations corresponding to rpscale = 1, and aerodynamic uncertainties described above.

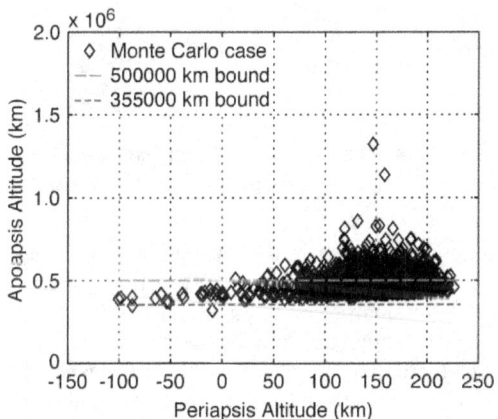

Figure 13a. Reference concept Monte Carlo results, apoapsis vs. periapsis. [8]

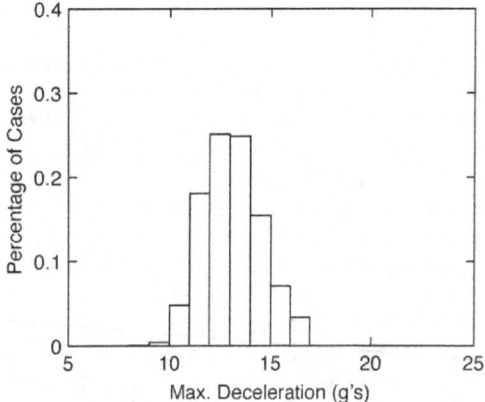

Figure 13c. Reference concept Monte Carlo results, heat load vs. peak heat rate. [8]

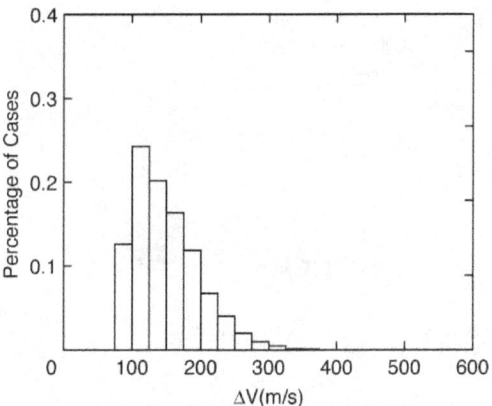

Figure 13b. Reference concept Monte Carlo results, delta V req'd to raise periapsis and correct apoapsis. [8]

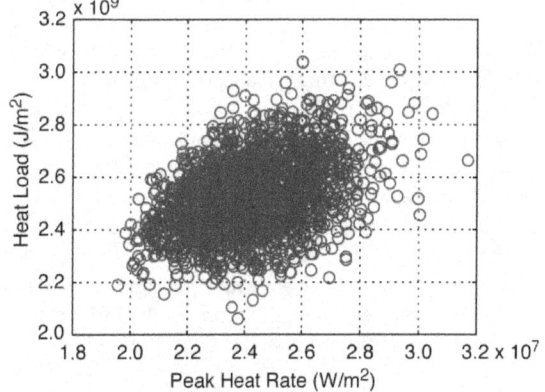

Figure 13d. Reference concept Monte Carlo results, heat load vs. peak heat rate. [8]

All 2001 Monte Carlo trajectories successfully capture. Approximately 66% of the cases achieve apoapsis within the apoapsis target range of 355,000 km and 500,000 km. In each of the Monte Carlo cases an apoapsis correction delta V, along with the periapsis raise delta V, is utilized to correct the orbit to the target of 430,000 by 3986 km. The total delta V, as shown in Fig. 13b, is 141 m/sec for the mean for the combined periapsis raise and apoapsis correction and 360 m/sec 99.87 percentile. Figures c and d illustrate the entry g loading and the peak heat rate vs. total heat load based on a stagnation point convective indicator for a 1m nose radius. The 3σ high g's are 17.6 g's, which are less than the 22.1 g's used to design the vehicle structure. The 3σ high heat rate and heat load stagnation point convective indicators are 2957 W/cm^2 and 295 kJ/cm^2, respectively, compared to the 3250 W/cm^2 and 290 kJ/cm^2 stagnation point convective rate and load indicators of the reference aeroheating design trajectory.

The apoapsis error (prior to delta V correction) for Neptune is greater than that seen in previous studies. Before any apoapsis correction, and a 430,000 km apoapsis target, the 3σ range in Neptune apoapsis is 371,300 to 832,700 km. For comparison the range in apoapsis at Titan,[1] prior to any delta V to adjust apoapsis, and a 1700 km apoapsis target, is 1499 km to 1883 km. The larger apoapsis errors at Neptune compared to Titan result from the high energy Neptune target orbit. At Neptune, the aerocapture exit velocity is very close to the Neptune escape velocity, resulting in a high sensitivity of apoapsis to aerocapture exit velocity. For example the ratio of the aerocapture exit velocity to escape velocity at Neptune is 0.97. The aerocapture to escape velocity at Titan is 0.69.[10]

The updated range of dispersions and uncertainties in navigation, aerodynamics, and atmosphere, are utilized in Fig. 13e, to assess the corridor margin for comparison to the original estimates in Fig. 5. The stacked aerodynamic uncertainties are used, and are similar to the assumptions earlier. Note that the atmosphere uncertainties are significantly less

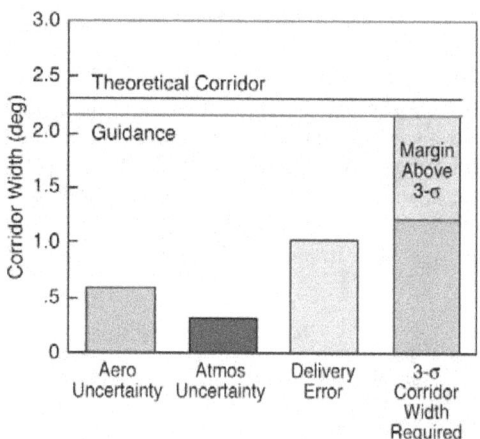

Figure 13e. Reference concept comparison of required vs. available theoretical corridor width.

than the initial estimates. This results from the reduced range of Fminmax, by incorporating the variation of density with latitude as opposed to utilizing a global range, and also due to the higher vehicle ballistic coefficient and reduced atmosphere variability at lower altitudes. The revised estimates show significant margin above the RSS value. The effects of high frequency density perturbations and additional aerodynamic uncertainties due to surface recession are not represented in the estimate, however. Results suggest that there may be margin in the performance design that could be utilized to reduce the vehicle L/D requirement, reduce the entry velocity or to accommodate increased atmosphere variability resulting from an increase vehicle size (and lower M/CDA) in an effort to reduce aeroheating.

ANGLE OF ATTACK MODULATION OPTION

Utilizing angle of attack control as an option to augment the bank angle modulation is considered to assess any potential benefits to performance and robustness.[10,8] Angle of attack modulation provides increased responsiveness to high frequency density perturbations and may assist with uncertainties in trim angle of attack. Angle of attack control could be provided with movement of an internal ballast or possibly with an aerodynamic control surface. Fig. 14a, b and c show results from the same Monte Carlo, one case without angle of attack control, and one case with ±5° angle of attack modulation. As shown, alpha modulation results in a significant reduction in apoapsis dispersions, delta V and g's.

AEROHEATING ENVIRONMENTS

The aeroheating design trajectory utilized for TPS sizing was based on the highest heat load trajectory from an earlier version of the reference Monte Carlo and is based on navi-

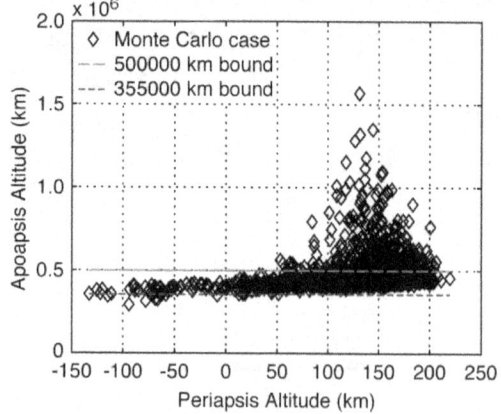

Figure 14a. Monte Carlo results without angle of attack modulation.[8]

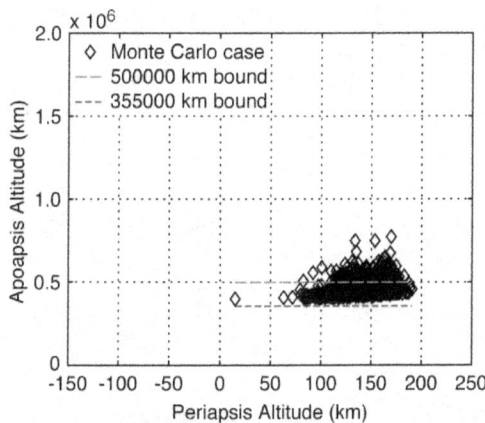

Figure 14b. Monte Carlo results with angle of attack modulation.[8]

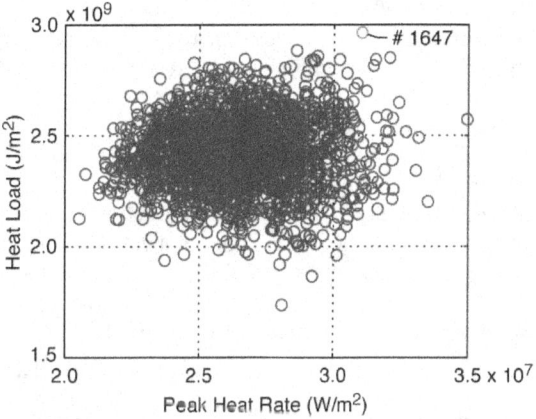

	Without α Modulation	With α Modulation
Apoapsis – 3σ high, low	12.85 E5, 3.25 E5	6.84 E5, 4.02 E5
Delta V – 3σ high, low	456 m/sec	288 m/sec
g's	20 g's	15 g's
Heat rate, load – 3σ high	3130 W/cm², 294 kJ/cm²	2968 W/cm², 277 kJ/cm²

Figure 14c. Comparison of performance parameters with and without angle of attack modulation.

gated uncertainties of ±.51° 3σ, the global range of Fminmax variability, rpscale = 1, and the aerodynamic uncertainties described earlier. Fig. 15a illustrates the range of heat rate vs. heat load for the Monte Carlo compared to the lift up lift down range. The Monte Carlo heat rate range is 2050-3250 W/cm², and heat load range is 195-290 kJ/cm². The lift up, lift down peak heat range is 3155-1122 W/cm², respectively. The lift up lift down heat load range is 185-442 kJ/cm², respectively. Typically the vehicle is designed to fly significantly closer to the center of the lift up lift down heat rate and load range than shown for these results. In this case, the guidance is designed to fly lift down early in the entry trajectory to allow successful targeting of the high-energy orbit apoapsis with the high ballistic coefficient vehicle. Fig. 15b illustrates the time variation of the stagnation point heating indicator for trajectory #1647 compared to that for the minimum atmosphere lift up and maximum atmosphere lift down trajectories. This further illustrates that the design and corresponding Monte Carlo results are skewed toward the lift up high heat rate profiles.

Because of these results, the peak heat load trajectory from the Monte Carlo, #1647, which also has ~98 percentile peak heat rate of 2001 trajectories, is selected as the reference trajectory for the TPS design, instead of the more traditional selection of the lift up trajectory for TPS selection, and lift down trajectory for TPS sizing.

Turbulent convective (LAURA and DPLR) and radiative (NEQAIR and RADEQUIL) computations are completed on the reference vehicle (m/C$_d$A = 895 kg/m² 2.88 m flattened ellipsled) lift up and lift down trajectories and are utilized to estimate "low", "med", and "high" aeroheating environments along Monte Carlo trajectory #1647.[11] Transition to turbulence prior to peak heating is

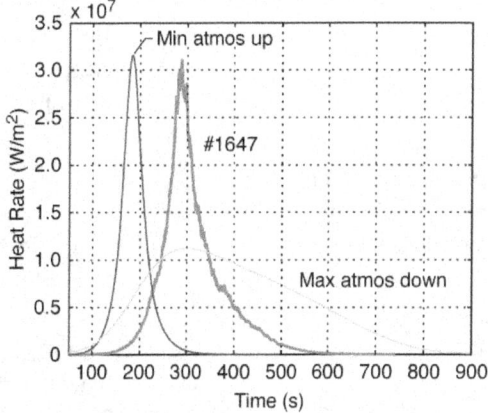

Figure 15a. Aeroheating design trajectory Monte Carlo results for convective stagnation point heat load vs. heat rate on a 1 m nose radius. Illustration of aeroheating design trajectory #1647.

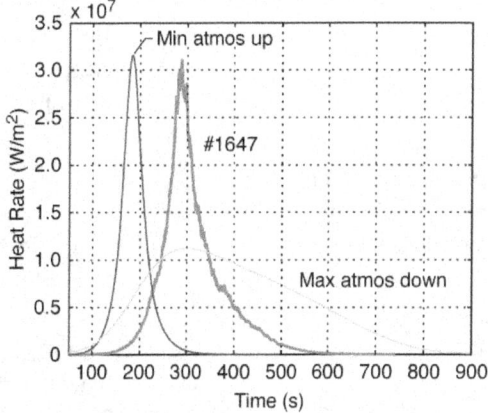

Figure 15b. Comparison of aeroheating profile for Monte Carlo trajectory #1647 to lift up minimum atmosphere and lift down maximum atmosphere trajectories.

expected due to significant ablation. Although only small differences, less than 10%, result in comparisons of LAURA and DPLR laminar aeroheating rates, large differences occur in turbulent heating comparisons between LAURA and DPLR. The turbulence models used in the analyses included Cebeci-Smith algebraic or Wilcox k-Ω model with LAURA; Baldwin-Lomax algebraic turbulence with DPLR. None of the turbulence models were developed for, or validated in, high Mach H2-He flows.

Radiation is a significant contributor to the Neptune aeroheating environments. Both NEQAIR and RADEQUIL are utilized to estimate the radiative aeroheating environments. Significant differences between the two predictions result. To assist in understanding the aeroheating environmnents, analyses of Galileo are completed using NEQAIR and RADEQUIL for comparison with historical analysis and flight data, and for comparison to the Neptune Orbiter study vehicle. Current uncoupled analyses predict the same order of magnitude results, (between 45.4 kW/cm2 and 78.5 kW/cm2 for the total uncoupled convection and radiation aeroheating) as the historical uncoupled analysis (63.3 kW/cm2 shown in Table 2). Engineering approximations, Galileo analysis and flight data indicate that the effects of convection/radiation/ablation coupling must be considered. No tools exist for modeling convection/radiation/ablation for coupled 3-D flowfields. (Galileo was modeled with 1-D assumptions.) Higher fidelity coupled models are expected to reduce the environments compared to uncoupled results. Development and validation of methods for modeling coupled convection/radiation/ablation 3-D flowfields is one of the technologies identified as enabling as a result of this study.

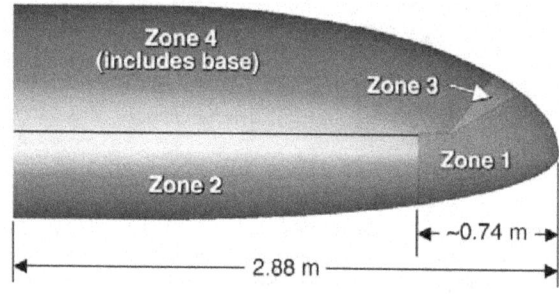

Figure 16. TPS zones.

Fig. 16 illustrates the division of the vehicle into zones, defined based on the vehicle structure, and to allow individual selection and sizing of TPS, based on the point with the highest rates and loads in each zone, to reduce overall TPS mass. Zone 1 and 2 comprise the heatshield or forebody of the vehicle. Zones 3 and 4 including the base, comprise the backshell. The vehicle maximum diameter, also referred to as the waterline, occurs at the boundary of zone 2 and 4.

Table 2. Comparison of Neptune Orbiter reference concept to Galileo.

	Galileo (Jupiter Dec 1995)	Neptune Orbiter (study)
Atmosphere composition	86.2% H2, 13.6% He	81% H2, 19% He
Inertial entry velocity (km/sec)	60	29
Atmos relative velocity (km/sec)	48	31.4
Inertial FPA (deg)	-6.835	−12.818
Trajectory	Ballistic	Lifting, guided, controlled
Configuration	44.25 deg sphere cone	Flattened ellipsled
Scale	1.25 m diam (.291 m nose rad)	2.88 m length
M/CDA (kg/m²)	224, 229	895
Heat pulse duration	~20 sec	~200 sec
Uncoupled stag pt peak heat rate (convec + radiative) (kW/cm²)	63.3	16
Coupled conv/rad/ablation (kW/cm²)	17.0 flight[12], 28.0 analysis[13]	??
TPS stagnation point thickness (cm)	14.6	12.9
TPS stagnation point recession (cm)	4.6	9.6
TPS material – heatshield	Nose piece: fabricated from billet of chopped molded carbon phenolic; tape-wrapped carbon phenolic flank	Nose: carbon phenolic (manufacturing approach??); Wind: reduced density carbon phenolic (dev/ testing?)

Figures 17a and b show the range "low", "med", and "high" of peak heat rate and load estimated based on the CFD and radiative aeroheating analysis for the highest heat rate location on both the vehicle nose and the vehicle

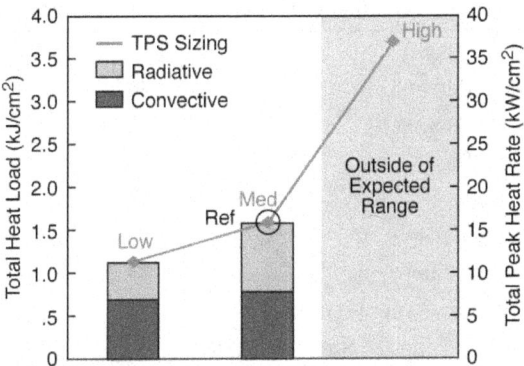

Figure 17a. Low, medium and high aeroheating results for zone 1.

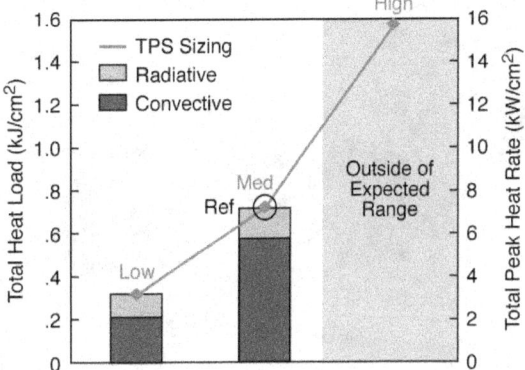

Figure 17b. Low, medium and high aeroheating results for zone 2.

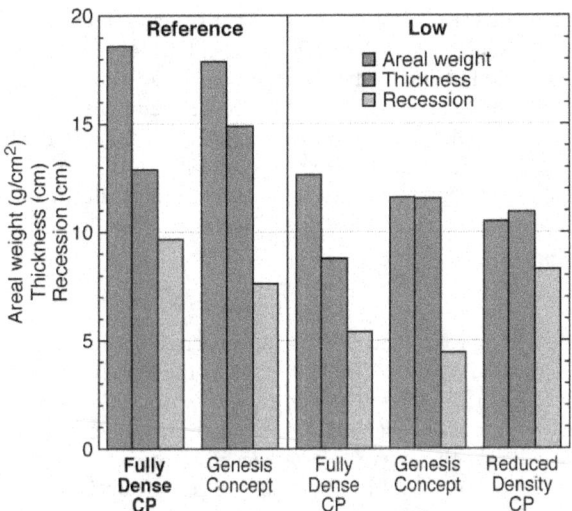

Figure 18a. TPS sizing results for zone 1. [14]

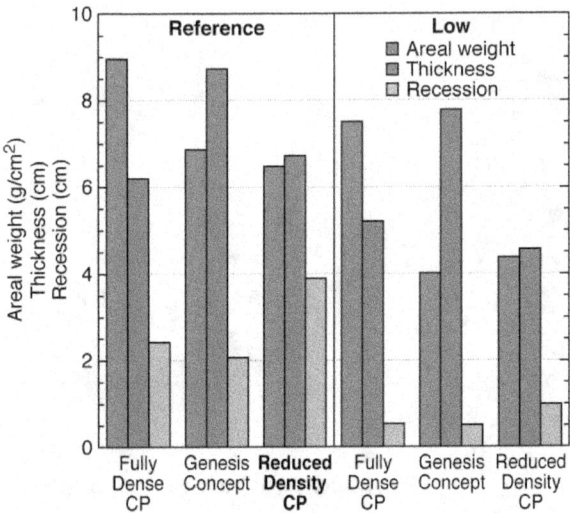

Figure 18b. TPS sizing results for zone 2. [14]

wind side for trajectory #1647. Note that after further analyses, including analysis of Galileo, the "high" estimate is well outside the expected range of aeroheating environments. These aeroheating environments are utilized to complete TPS selection and sizing.

TPS materials considered for the heatshield include carbon phenolic, reduced density carbon phenolic, and the Genesis carbon fiber form with carbon-carbon face sheet concept. Results of TPS sizing for the "Low" and "Medium" aeroheating are completed and shown in Fig. 18a and b, for the nose and wind side, respectively. [14] "Medium" levels are utilized for the Reference. The nose region is characterized by significant recession. Fabrication of the tape-wrapped carbon phenolic or Genesis concept may not be possible for these environments. As a result a fully dense carbon phenolic is selected for the nose region of the reference concept. However, TPS thickness in the nose region is beyond current TPS manufacturing experience for this shape and acreage. If the aeroheating rates and loads remain at the levels estimated, TPS manufacturing approaches will be enabling for the Neptune aerocapture mission. For the wind side, the reduced density carbon phenolic is selected, but additional work is needed to design and assess the ability of this type of concept to accommodate the heat rates estimated.

Zone	Material	Mass (kg)
Zone 1 (Nose)	Fully Dense CP	204
Zone 2 (Wind)	Reduced Density CP	293
Zone 3 (Lee, Nose)	PICA	0.6
Zone 4 (Lee, Nose)	SLA 561	58

Figure 18c. Reference concept TPS selected and corresponding CBE mass. [14]

AEROSHELL STRUCTURE

Fig. 19a and b show the reference vehicle structural

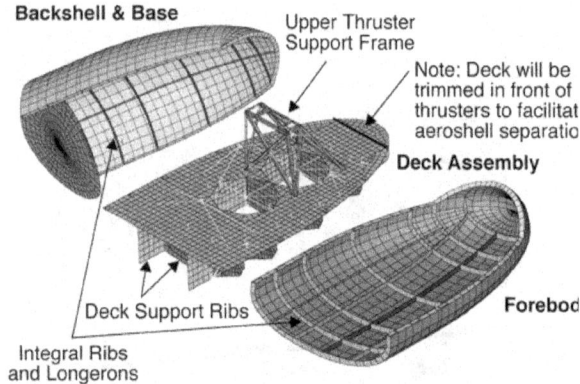

Component	CBE Mass (kg)	
Aeroshell	94.85	
Forebody	44.9	
Backshell	42.8	
Base	7.15	
Deck		21.6
Deck Ribs		17.9
Tank Supt Rods		.42
Thruster Supt Frame		1.75
Total Ellipsled Structure (CBE)		136.5

Figure 19a. Reference structural concept. [15]   **Figure 19b. Reference concept CBE masses.** [15]

concept and initial current best estimate of mass properties, respectively.[15] Optimization of the structure after completion of the study indicates an opportunity to reduce the structural mass.[15] Launch loads and stiffness requirements, and aerocapture entry loads are considered in the design and sizing of the structure. The load path for the orbiter on launch is from the cruise stage through the aeroshell to the deck. The TPS mass is considered to be a parasitic mass. The aeroshell forebody, backshell, base and deck are 2.54 cm thick sandwhich construction with 5052 Al honecomb core and Graphite/Polymide face sheets. Integral monolithic blade stiffeners, longerons and ribs, are included for the forebody and backshell. The deck includes deck support ribs. 20 separation fittings attach the aeroshell forebody and backshell, and deck, which are used to separate the backshell and forebody from the deck after aerocapture.

MASS PROPERTIES, PACKAGING
Fig. 20a, b illustrate the packaging of the aerocapture orbiter, two probes and SEP propulsion module in the 5m Delta IVH fairing[3]. Fig. 20c illustrates the packaging of the aerocapture orbiter. Table 3 includes the mass summary of the reference vehicle concept. The stack wet launch allocation is 5500kg. The aerocapture entry allocation is 2238kg (~2% greater than the allocation used in the performance analyses). 35% margin (allocation – CBE)/allocation is included on dry mass, with ~8% unallocated launch reserve. The aerocapture mass fraction is 59% of the orbiter dry mass based on growth masses ("MEV" in Table 3) with aerocapture propellant included (aeromaneuvering, periapsis raise and apoapsis correction); and 50% without aerocapture propellant included.

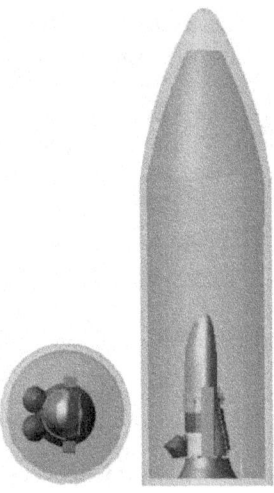

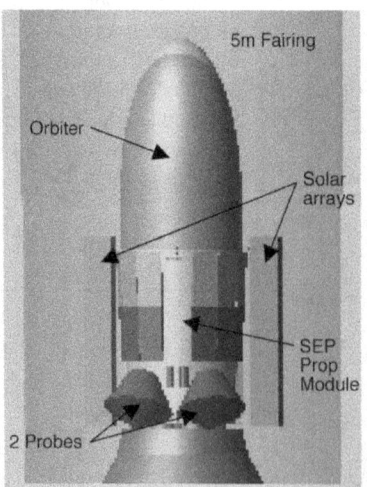

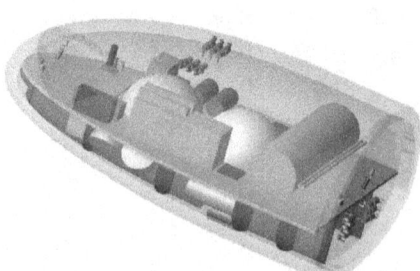

Figure 20a. Reference concept packaging in Delta IV, 5 m fairing. [3]   **Figure 20b. Detail of reference concept packaging in Delta IV, 5 m fairing.** [3]   **Figure 20c. Reference concept orbiter packaging.** [3]

Table 3. Reference concept mass property summary.[3]

Mass in kg	CBE	Cont	MEV	Marg	Alloc
Launch Capability					**5964**
Launch Reserve				8.4%	463
Launch Wet Alloc					**5500**
SEP LV Adapter	48	30.0%	62	12.2%	70
Xenon	973	10.0%	1070	0.0%	1070
SEP Dry Mass	1134	29.5%	1468	20.0%	1762
Cruise Hydrazine			111		111
Cruise Probes	159	30.0%	207	20.0%	249
A/C Entry Alloc					**2238**
A/C Aeroshell/TPS	736	30.0%	957	20.0%	1149
A/C ACS Prop			22		22
A/C Peri Raise Prop			139		139
Orbit Wet Alloc					**928**
Orbit Prop			124		124
Orbit Dry Mass	524	27.3%	667	20.4%	804

CBE = Current Best Estimate
Cont = Contingency = (MEV-CBE)/CBE
MEV = Maximum Expected Value
Marg = Margin = (Alloc-MEV)/MEV
Alloc = Allocation

COMPARISON TO ALL-PROPULSIVE MISSION

Several alternate mission concepts are shown in Table 4[4] for comparison to the reference concept labeled "Option B2". Each option shows the mass that can be delivered to Neptune prior to insertion, labeled "Pre-NOI Net Delivered Mass.," and the mass required to capture into Neptune orbit. For the chemical insertion the chemical propellant and chemical dry mass are calculated based on the "Pre-NOI Net Delivered Mass". For the aerocapture system, the "Aerocapture System" mass is based on the reference concept and is fixed at 1119 kg. The "Payload in Neptune Orbit" is defined based on the reference concept and is 792 kg. "System Margin" represents either a surplus or deficit in the capability of the system to deliver the 792 kg into orbit. The System Margin should be between 15-20% for adequate margin.

To determine the benefit of aerocapture compared to an all-propulsive system, the aerocapture system that delivers the maximum mass to Neptune orbit (Delta IVH, EJGA, SEP, Aero) can be compared to the all-propulsive system that delivers the maximum mass to Neptune orbit (Delta IVH, EJGA, SEP, Chem), each for the same launch vehicle. For the all-propulsive option, a maximum of 1167 kg can be delivered into Neptune orbit (zero margin). For the aerocapture option, assuming a fixed aerocapture mass fraction of 59% (includes aerocapture deltaV), 1614 kg can be delivered into Neptune orbit (zero margin). Therefore aerocapture results in approximately 1.4 times more mass in Neptune orbit as compared to an all-propulsive system.

In addition, Table 4 shows significant trip time savings for the aerocapture systems as compared to the all-propulsive systems.

Table 4. Comparison to alternate mission concepts. [4]

Launch Vehicle	Delta IV H							Atlas 551	
Gravity Assist	VEJGA	EJGA				VJGA		EJGA	
Earth to Neptune Prop System	Chem	Chem		SEP		SEP		Chem	SEP
NOI Prop System	Chem	Aero	Aero	Chem	Aero	Chem	Aero	Aero	Aero
Option	A1	A2	A2	B1	B2	B1	B2	A2	B2
							[7]		
Cruise Time to Neptune (yrs)	15.0	10.8	11.8	15.0	10.5	15.0	10.3	11.8	10.5
Launch Year	2014	2016	2014	2016	2016	2017	2017	2014	2016
Launch C3 (km2/sec2)	15.6	26.0	47.3	13.5	13.6	17.0	18.4	47.3	9.1
SEP Power (kW, EOL)				30	30	30	30		30
Inertial Entry Velocity (km/s)		29	29		29		29	29	29
Neptune Cruise Chem DV (m/s)[1]	3429	1413	357					357	
NOI Chem DV (m/s)[1]	2300			2871		2781			
	[9]			[6]		[6]		[6]	
Launch Capability	7012	5695	3550	6543	6532	6130	5964	2630	4850
Propellant Mass[2,3]	4158	2040	376	655	809	1025	1070	279	713
LV to Prop Module Adapter	62	62	62	62	62	62	62	62	62
Prop Module Dry Mass	806	542	289	1437	1449	1465	1468	243	1441
Chem Prop Mod to Payload Adapter	40	40	40					40	
Pre-NOI Separated Mass[10]	318	318	318	318	318	318	318	318	318
Pre-NOI Net Delivered Mass	1628	2694	2464	4071	3895	3260	3046	1688	2315
Aerocapture System[4]		1119	1119		1119		1119	1119	1119
NOI Chem Propellant Mass[8]	966			2417		1898			
NOI Chem Dry Mass	280			487		413			
Payload in Neptune Orbit	792	792	792	792	792	792	792	792	792
System Margin = LV-MEV	(409)	783	553	375	1984	157	1135	(223)	404
System Margin % = (LV-MEV)/MEV	-5.5%	15.9%	18.5%	6.1%	43.6%	2.6%	23.5%	-7.8%	9.1%

MEV: Maximum Expected Value = best estimate + 30% contingency
Assumptions and Notes:
All masses are MEV mass listed in kg
[1] Includes 5% DV contingency
[2] Chem Propellant mass calculated using "Launch Capability" as system total mass; Chem Isp = 325 sec
[3] SEP Propellant mass calculated using "Launch Capability" as system total mass; includes 10% prop mass contingency
[4] Aerocapture System Mass: aeroshell structure, TPS, and DV to achieve 28766x488,000 km orbit
[6] Propellant mass and Prop Module Dry Mass for SEP / Chem options includes propellant and dry mass for
    both SEP and chemical stages
[7] Neptune Aerocapture Study Reference Mission
[8] Chem Propellant mass calculated using "Pre-NOI Net Delivered Mass" as Initial mass; Chem Isp = 325
[9] Total Cruise+NOI DV split equally between two stages; I.e. Cruise delta-V is staged
[10] Includes Probes and ~100kg of cruise hydrazine

SUMMARY AND TECHNOLOGY

Aerocapture can deliver 1.4 times more mass to Neptune than an all-propulsive system for the same launch vehicle. Aerocapture is feasible and performance is adequate for the Neptune aerocapture mission. Monte Carlo simulation results show 100% success for all cases including conservative assumptions on atmosphere and navigation. Additional analyses are required to assess the amount of surface recession from coupled 3-D convective/radiative/ablation analyses, determine the aerodynamics and uncertainties resulting from time and path dependent shape change, and evaluate the effect on guidance and control algorithm design, and performance. The Neptune spacecraft can be successfully packaged in an aeroshell and result in ~8% unallocated mass while meeting the required mass margins.

Technologies identified in the study as requiring development are grouped into three categories; enabling technologies, strongly enhancing technologies and enhancing technologies. Technologies annotated with an asterisk are categorized based on current understanding. Additional assessment could change the categories.

The enabling technologies identified include

- TPS Manufacturing. TPS thicknesses are beyond current manufacturing experience for carbon phenolic for this shape and acreage.
- Aerothermodynamic methods and validation

o Aerothermodynamics are characterized by high radiative and convective aeroheating, coupled convection/radiation/ablation, and significant surface recession with effects on vehicle aerodynamics on a more complex shape.

o Coupled convection/radiation/ablation capability for three-dimensional flowfields is needed for definition of aeroheating environments, TPS requirements, and vehicle shape change.

o An approach is needed to determine and represent the aerodynamics/uncertainties on the time varying path dependent shapes and corresponding masses in an aerodatabase and simulation.

The strongly enhancing technologies identified include

- Guidance Algorithm* – Existing guidance algorithms have been demonstrated to provide adequate performance. However, improvements are possible to improve performance, to determine the ability to reduce heat loads and to accommodate time varying, path dependent shape and ballistic coefficient change
- Flight Control Algorithm* – Algorithms must be able to accommodate shape change uncertainties
- Atmosphere Modeling – Neptune General Circulation Model output is needed to represent the dynamic variability of the atmosphere.
- Reduced Mass TPS concepts, ex., reduced density carbon phenolic, could be utilized to decrease aeroshell mass.
- Utilizing the TPS as a structural element may reduce the combined structure plus TPS mass.
- Alpha Modulation* reduces the dispersions in apoapsis, provides additional and more rapid response to density perturbations, and provides additional margin for trim angle of attack uncertainties.
- Dual Stage MMRTGs
- Deployable Ka-Band HGA

The enhancing technologies identified include

- Automated navigation, improved optical navigation camera.
- Miniaturized ACS components.
- Lower Mass, Power Science Instruments

FUTURE WORK

Several areas are recommended for future systems analysis in addition to the specific technology items listed above. Recommendations are as follows.

Complete partial design cycles for one or more intermediate (between 2.88 m-5.5 m length scale) vehicle sizes. There may be a minimum mass vehicle between the 2.88 m and 5.5 m length vehicle. The trade is surface area vs. areal density of the combined TPS and structure.

The current design has 460 kg unallocated mass. In addition, interplanetary trajectory designs have resulted in increased delivered mass capability. Several design changes can be considered within the increased mass capability. For example, an increased vehicle scale may reduce aeroheating rates and loads and the corresponding surface recession and TPS thickness required.

Additional trades that can be completed include a further assessment of chemical vs. SEP cruise; additional systems analysis considering angle of attack modulation; revisiting the L/D=.6 vehicle; consideration of a symmetric version of the flattened ellipsled i.e. an elliptic upper section, in addition to elliptic lower and section; utilization of the TPS as a structural element; and consideration of variable thickness TPS for TPS mass reduction.**ACKNOWLEDGMENTS**

The author would like to acknowledge and thank the team members of the NASA Aerocapture Systems Analysis Team at ARC, JPL, JSC, LaRC and MSFC for their work and contributions to the Neptune Aerocapture Systems Analysis Study and to this paper. Thank you to Paul Wercinski, Aerocapture Systems Analysis Study Peer Review Chair, and the peer review panel members for review of this work and helpful comments and recommendations. Thank you to Code S In Space for sponsoring this work. Thank you to Anne Costa for preparing this paper for publication.

REFERENCES

[1]Lockwood, M.K., "Titan Aerocapture Systems Analysis," AIAA-2003-4799, July, 2003.
[2]"Neptune Orbiter/Probes 2001-12" TeamX 6,7,14 December 2001.

[3]Bailey, R.W., Hall, J.L., Spilker, T.R., O'Kongo, N., "Neptune Aerocapture Mission and Spacecraft Design Overview," AIAA-2004-3842, July 11-14, 2004.

[4]Noca, M., Bailey, R.W., "Mission Trads for Aerocapture at Neptune," AIAA-2004-3843, July 11-14, 2004.

[5]Edquist, K.T., Prabhu, R.K., Hoffman, D.A., and Rea, J.R., "Configuration, Aerodynamics and Stability Analysis for a Neptune Aerocapture Orbiter," AIAA-2004-4953, August 16-19, 2004.

[6]Haw, R. "Aerocapture Navigation at Neptune," AAS-03-643, August 3-7, 2003.

[7]Justus, C.G., Duvall, A., Keller, V.W. "Atmospheric Models for Aerocapture Systems Studies," AIAA-2004-4952, August 16-19, 2004.

[8]Starr, B.R., Powell, R.W., "Aerocapture Performance Analysis for a Neptune- Triton Exploration Mission," AIAA-2004-4955, August 16-19, 2004.

[9]Cobleigh, B.R., "Development of the X-33 Aerodynamic Uncertainty Model," NASA TP-1998-206544, April 1998.

[10]Masciarelli, J., Westhelle, C.H., Graves, C.A., "Aerocapture Guidance Performance for the Neptune Orbiter," AIAA-2004-4954, August 16-19, 2004.

[11]Hollis, B., Takashima, N., Sutton, K., Wright, M., Olejniczak, J., Prabhu, D., "Preliminary Convective-Radiative Heating Environments for a Neptune Aerocapture Mission," AIAA- 2004-5177, August 16-19, 2004.

[12]Tauber, M.E., NASA TM-1999-208796, Sep. 1999

[13]Moss, J.N., Dimmonds, A.L., AIAA 82-0874

[14]Laub, B., Chen, Y.-K., "TPS Challenges for Neptune Aerocapture," AIAA-2004-5178, August 16-19, 2004.

[15]Dyke, R.E., Hrinda, G., "Structural Design for a Neptune Aerocapture Mission," AIAA-2004-5179, August 16-19, 2004.

NEPTUNE AEROCAPTURE MISSION AND SPACECRAFT DESIGN OVERVIEW

R. W. Bailey, J. L. Hall, T. R. Spilker, N. O'Kongo
Jet Propulsion Laboratory, California Institute of Technology, Pasadena, CA, 91109

A detailed Neptune aerocapture systems analysis and spacecraft design study was performed as part of NASA's In-Space Propulsion Program. The primary objective was to assess the feasibility of a spacecraft point design for a Neptune/Triton science mission that uses aerocapture as the Neptune orbit insertion mechanism. This paper provides an overview of the science, mission and spacecraft design resulting from that study. The estimated delivered wet mass allocation to Neptune orbit was ~928 kg. The aerocapture entry system, comprised of aeroshell and post-aerocapture orbit correction propellant, was ~1252 kg, for a total atmospheric entry mass allocation of ~2239 kg. The aeroshell used was a 2.88 m long flattened ellipsled with a lift to drag ratio of 0.8. A Delta-IV Heavy launch vehicle combined with a 30kW solar electric propulsion (SEP) stage and a Venus/Jupiter gravity assist were used to get the spacecraft to Neptune in 10.25 years. The SEP stage and Orbiter both have 35% dry mass margins ((allocation – CBE)/allocation) and the overall launch stack has an additional ~8% unallocated reserve. The feasibility of the mission requires the solution of two key technical challenges: improvement in aerothermodynamic computational tools for Neptune; and development of thermal protection material manufacturing processes for the increased thickness needed for aerocapture. Several other component technologies were identified as being able to provide significant performance improvements including: radioiostopic power generation, solar cells and array structure, low mass/power science instruments, and small stowed volume/large aperture deployable Ka-Band antennas.

NOMENCLATURE

ACS	=	Articulation and Attitude Control System	*L/D*	=	Lift over Drag
Alloc	=	Allocation	*LGA*	=	Low Gain Antenna
AU	=	Astronomical Unit	*LHP*	=	Loop Heat Pipe
CBE	=	Current Best Estimate	*LV*	=	Launch Vehicle
C&DS	=	Command and Data System	*Marg*	=	Margin: (Alloc – MEV) / MEV
Cont	=	Contingency: (MEV – CBE) / CBE	*MEV*	=	Maximum Expected Value
CM	=	Center of Mass	*MMRTG*	=	Multi-Mission Radioisotpic Thermal Generator
DSN	=	Deep Space Network	*PPU*	=	Power Processor Unit
dV	=	Delta Velocity	*RF*	=	Radio Frequency
EMI/EMC	=	Electromotive Interference/ Electromotive Compatibility	*RTG*	=	Radioisotpic Thermal Generator
EOL	=	End of Life	*SEP*	=	Solar Electric Propulsion
EOM	=	End of Mission	*SSPA*	=	Solid State Power Amplifier
Gbits	=	Gigabits	*TCM*	=	Trajectory Correction Maneuver
HGA	=	High Gain Antenna	*TPS*	=	Thermal Protection System
IR	=	Infrared	*TRL*	=	Technology Readiness Level
JPL	=	Jet Propulsion Laboratory	*TWTA*	=	Traveling Wave Tube Amplifier
kg	=	kilograms	*UHF*	=	Ultra High Frequency
km	=	kilometers	*UV*	=	Ultraviolet
kW	=	kilowatt	*W*	=	Watts

INTRODUCTION

Aerocapture is being investigated as a means for interplanetary orbit insertion by NASA's In Space Propulsion Program. A systems analysis and spacecraft point design study was performed in Fiscal Year 2003 based on a reference mission to Neptune. The purpose of this study was to quantify the feasibility and performance of an aerocapture system to insert a spacecraft into a scientifically useful orbit about Neptune that includes regular flybys of Neptune's moon Triton. This paper is one of eleven papers (Ref 1-10) associated with this Neptune aerocapture study; see Ref 1 for an overview of the entire Neptune aerocapture study. The multi-center Neptune Aerocapture team is largely the same team that performed a similar mission study for a Titan Aerocapture in 2002[11,12].

Of the twelve study papers, this paper discusses the science objectives and the resulting spacecraft configuration. Disciplines addressed in the other papers (Ref 2-10) include: trajectory design[2], deep space navigation[3], Neptune atmospheric models[4], aerodynamics and stability analysis[5], aerocapture guidance[6] and performance analysis[7], Neptune radiative and convective heating environments[8], TPS[9], and structural design/analysis[10].

SCIENCE

Science objectives for a Neptune/Triton mission were selected to yield a payload that representatively stressed spacecraft accommodation issues. The payload used for the study does not constitute a fully vetted recommendation for a Neptune/Triton mission, but instead represents a typical range of requirements such a mission might expect.

SCIENCE OBJECTIVES & MEASUREMENTS

Science objectives were composed from various sources including:

- 2002 Solar System Exploration Decadal Survey
- Science community White Papers that contributed to the SSEDS and Individual scientists that were on the SSEDS Panels
- NASA SSE and SEC Roadmaps, 2003

The science objectives used for this study were (listed in no particular order):

1. Global imaging spectrometry of Neptune, Triton, other satellites, and the rings, at UV, visible, and IR wavelengths; repeated as needed for time variability
2. Global microwave radiometry at Neptune to infer temperature and pressure as a function of altitude
3. Measure the low-order gravity fields of Neptune satellites; at Triton, measure higher-order harmonics to infer gross interior structure
4. Measure Neptune's magnetic field at low altitudes, with sufficient accuracy and spatial and temporal coverage to map its generation region and determine its temporal variability
5. Measure abundances of key atmospheric constituents as a function of depth (Neptune)
6. Measure atmospheric temperature and pressure as a function of depth (Neptune)
7. Measure winds as a function of depth (Neptune)
8. Microwave radiometry to infer temperature and pressure as a function of depth (Neptune)
9. Measure the energetic charged particle environment
10. Measure the plasma wave environment

SCIENCE INSTRUMENTS

Table 1 presents 1) the instrument suite selected to achieve the science objectives, and 2) the spacecraft accommodation considerations for the instruments. The mass and power for many of these instruments would be considered aggressive for a mission in formulation today. Instrument development would be required to achieve the payload capability for the total mass/power allocation presented.

Table 1. Science Instruments

Instrument	Spatial Res Neptune (km)	Triton (km)	Global Coverage (Neptune & Triton)	# of channels/ spectra/ wavelens/ samples	Bits per channel per spatial resolution	Total Data (Gbits)	Mass (kg)	Avg Power (W)
Visible Imager (low-res)	100	10	100%	7	10	28.03	5.2	3
(high-res)	10	1	10%	7	10			
Infrared Imaging Spectrometer (low-res)	1000	100	100%	1024	8	32.88	5.85	10
(high-res)	100	10	10%	1024	8			
Thermal IR Imaging Spectrometer (low-res)	5000	500	100%	50	8	0.06	4.2	1
(high-res)	500	50	10%	50	8			
Ultra Violet Imaging Spectrometer (low-res)	1000	100	100%	1024	8	32.88	2.0	3
(high-res)	100	10	10%	1024	8			
	Integrate Time (s)		Duration (s)					
Microwave Radiometer	0.1		43200	6	12	0.62	5.2	10
Magnetometer	0.1		2592000	3	24	3.73	0.8	1
UVIS Occultation Port	0.01		36000	512	8	29.50	1.3	3
Mass Spectrometer	60.0		72000	1	50000	0.12	6.5	6
Charged Particle Detector			Allocations			1	7.8	2
Plasma Wave Spectrometer						22	5.2	10

MISSION OVERVIEW

Certain aspects of the mission were assumed as ground rules. Other aspects of the mission were open to system trades and/or inherited from other outer planet mission studies performed at JPL.

GROUND RULES

Several ground rules and assumptions were set to bound the study. These items were not subject to any system trades analysis.

- The TRL 6 cutoff date shall be no later than 2010.
- The Neptune atmospheric probe will be a "black box" with a 124 kg launch mass allocation (each).
- The orbiter shall perform an aerocapture for Neptune orbit insertion.
- The Earth to Neptune propulsion system shall be SEP.
- The orbiter shall accommodate the instruments in Table 1 for their intended science purpose.

EARTH TO NEPTUNE TRAJECTORY

The Earth to Neptune SEP trajectory selected for this study, shown in Figure 1, was the result of an extensive analysis of Chemical / SEP cruise and chemical/aerocapture Neptune insertion options (Ref 2). This trajectory provides a Neptune atmospheric entry velocity of 29 km/s.

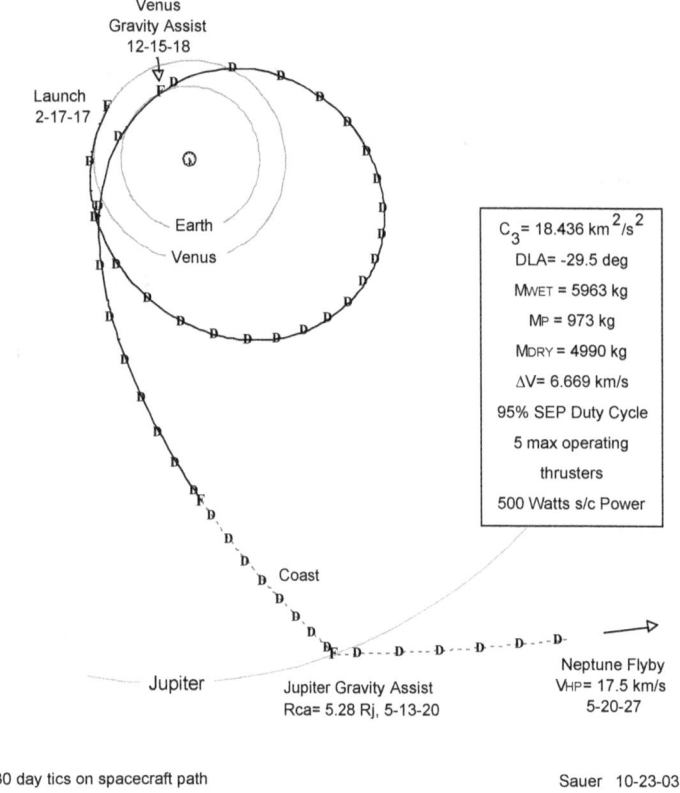

10.25 Year VJGA Neptune Flyby
Delta IV (4050H-19) / SEP 30 kW

Venus Gravity Assist 12-15-18

Launch 2-17-17

Earth

Venus

$C_3 = 18.436 \text{ km}^2/\text{s}^2$

DLA = -29.5 deg

$M_{WET} = 5963$ kg

$M_P = 973$ kg

$M_{DRY} = 4990$ kg

$\Delta V = 6.669$ km/s

95% SEP Duty Cycle

5 max operating thrusters

500 Watts s/c Power

Coast

Jupiter

Jupiter Gravity Assist
Rca= 5.28 Rj, 5-13-20

Neptune Flyby
$V_{HP} = 17.5$ km/s
5-20-27

30 day tics on spacecraft path

Sauer 10-23-03
vjga17-31-120x

Figure 1. Earth to Neptune Trajectory

MISSION TIMELINE

The mission timeline is listed below. For the "Time" column, 'L' = Launch, 'A' = Orbiter atmospheric interface, 'y' = years, 'd' = days, 'h' = hours, and 'm' = minutes.

Time	Event
L+0	Launch, SEP burn start
L+10m	Venus flyby
L+32m	SEP burn out, solar array jettison
A-135d	TCM 1
A-121d	Probe 1 Release TCM (2)
A-120d	Probe 1 Release and separate
A-114d	Probe 2 Release TCM (3)
A-113d	Probe 2 Release and separate
A-105d	Atmospheric Interface Target TCM (4)
A-60d	TCM 5
A-10d	TCM 6
A-36h	TCM 7 (if needed)
A-3.5h	Probe 1 entry
A-2h	Probe 1 end of life (EOL)
A-2h	Probe 2 entry
A-30m	Probe 2 EOL
A-30m	Jettison non-aero external components
A-29m	Align for aerocapture interface
A+38h	Jettison aeroshell
A+77h	Periapsis raise burn
A+2y	End of Mission

The probes are released 90 minutes apart to allow the first probe to reach its end of mission before the second probe enters the atmosphere.

The Neptune approach trajectory and subsequent aerocapture flight is retrograde to Neptune (157 deg inclination) to match Triton's orbit plane and motion. The primary deceleration pulse of aerocapture lasts less than 10 minutes, during which the Orbiter is modulating its trajectory with bank angle using bi-propellant thrusters. The aeroshell is jettisoned soon after atmospheric exit to limit thermal soak back from the heatshield to the orbiter. The orbiter then performs a TCM at the first post-aerocapture apoapsis (~430000km) to raise the periapsis out of the atmosphere and up to the desired science altitude (~4000 km). The final desired science orbit is 4000x488000km altitude @ 157 deg inclination (Triton's orbit is ~circular at 330000km altitude). The science orbit provides a Triton flyby every 3 orbits (~11.75 days).

MISSION SYSTEM DESCRIPTION

The Neptune Orbiter Flight System was the primary focus of this study; Ground Data and Mission Operations Systems were out of scope of this study. The launch configuration is shown in Figure 2. The stack fits easily inside the Delta IV fairing and looks slightly off-center geometrically because of the center of mass alignment within the aeroshell for proper aeroshell angle of attack and stability. The launch system mass summary is shown in Table 2. The SEP cruise, post-SEP cruise, and probe communications relay configurations are shown in Figures 3-5 respectively.

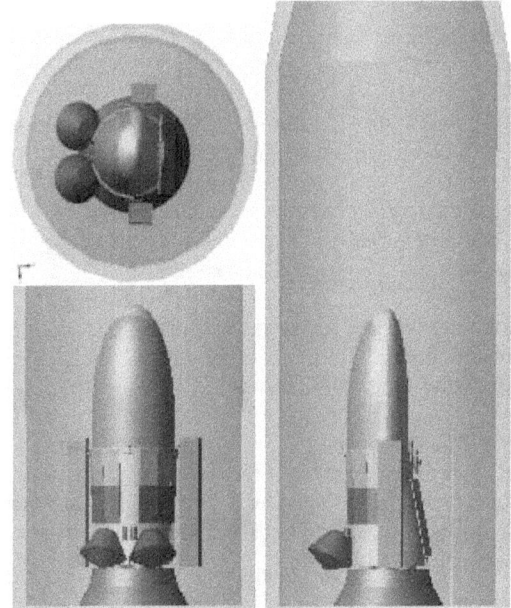

Figure 2. Delta IV H Launch Configuration

Table 2. Launch Mass Summary

Mass in kg	CBE	Cont	MEV	Marg	Alloc
Launch Capability					**5964**
Launch Reserve				8.4%	463
Launch Wet Alloc					**5500**
SEP LV Adapter	48	30.0%	62	12.2%	70
Xenon	973	10.0%	1070	0.0%	1070
SEP Dry Mass	1134	29.5%	1468	20.0%	1762
Cruise Hydrazine			111		111
Cruise Probes	159	30.0%	207	20.0%	249
A/C Entry Alloc					**2238**
A/C Aeroshell/TPS	736	30.0%	957	20.0%	1149
A/C ACS Prop			22		22
A/C Peri Raise Prop			139		139
Orbit Wet Alloc					**928**
Orbit Prop			124		124
Orbit Dry Mass	524	27.3%	667	20.4%	804

CBE = Current Best Estimate
Cont = Contingency = (MEV-CBE)/CBE
MEV = Maximum Expected Value
Marg = Margin = (Alloc-MEV)/MEV
Alloc = Allocation

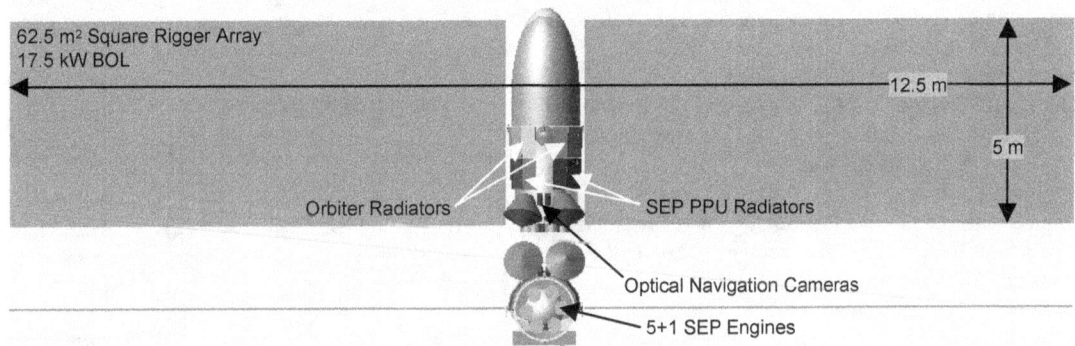

Figure 3. SEP Cruise Configuration

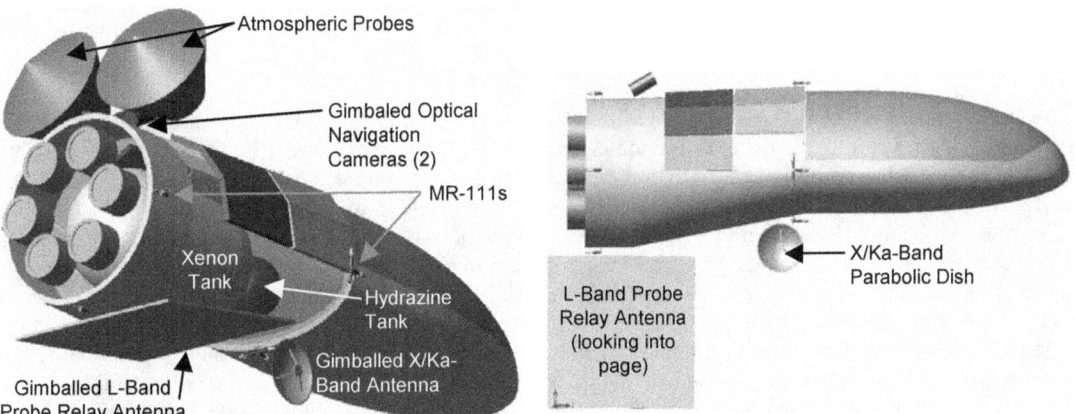

Figure 4. Post SEP Cruise Configuration

Figure 5. Probe Comm Relay Configuration

KEY MISSION SYSTEM TRADES

Several trades associated with how the orbiter interacts with the rest of mission system are worth mentioning. These trades do not represent a complete trade space for the Neptune mission.

1. Number of Probes and Probe Delivery

The Probes could be delivered before or after aerocapture. The volume constraints of the aeroshell led to the probes being carried outside the aeroshell, thus delivery prior to aerocapture. Three probes were desired at the start of the study. Although three probes could probably be accommodated on the launch vehicle, the relay telecom solution could not converge because the first probe would be too far away to provide a meaningful data rate. Three probes at one hour each versus two probes at 1.5 hours each could be a reasonable solution, but the probes would require higher ballistic coefficients which could complicate packaging.

2. Probe Lifetime

Because the probes are released prior to aerocapture, and the orbiter reaches Neptune with a speed of 29 km/s, probe lifetime equates directly to telecomm relay distance for the probe. 90 minutes was selected, along with the largest non-deployable L-Band antenna configuration, to achieve relay for two probes relayed in sequence.

3. Neptune/Triton Science Orbit

The desire to insert into an orbit that resonates with Triton's led to an aerocapture exit apoapsis that was so high (488,000 km) that the margin between aerocapture to that apoapsis and escape from the Neptune gravity well was too close to the performance capability of the aerocapture system. To achieve a comfortable aerocapture performance margin, the aerocapture exit apoapsis altitude was lowered to 430,000 km and propellant was added to the orbiter to allow the orbiter to raise the apoapsis to the desired science orbit.

SEP STAGE DESIGN

The SEP stage is designed to provide mission functionality from launch through Probe 2 EOL. The SEP stage relies on the orbiter for its flight computer and attitude control (reaction wheels). Structural mass was parametrically scaled against the orbiter/aeroshell mass (payload), Xenon mass, and other primary components. The SEP stage dry mass summary is shown in Table 3.

The "Flt" column of Table 3 specifies the number of line items in the detailed mass list for the respective subsystem. Articulation and Att Control includes gimbals and actuators for SEP thrusters, cameras, and antennas. Attitude control for the system is provided by the reaction wheels in the orbiter. The hydrazine propulsion system provides TCM capability and momentum de-saturation. For Telecomm, pre-aerocapture antennas and L-Band Probe relay radios are included; all X/Ka band radio equipment resides in the orbiter. The Power system is mostly solar arrays. For Thermal, all the SEP PPU radiators have louvers.

Table 3. SEP Stage Dry Mass Summary

Mass in kg	Flt	CBE	Cont	MEV
SEP Dry Mass	193	1133.8	29.5%	1468.3
Articulation & Att Control	16	46.4	25.2%	58.1
Telecom	11	16.4	30.0%	21.3
Power	5	319.2	29.7%	413.8
Propulsion	25	288.5	30.0%	375.1
Structure	32	310.8	30.0%	404.0
Cabling	28	69.1	30.0%	89.8
Hydrazine Propulsion	50	19.0	17.9%	22.4
Thermal	26	64.5	30.0%	83.9

ORBITER DESIGN

The Orbiter Flight System (OFS) is a single fault tolerant system, except for structure, dual stage RTGs, and antennas, with dual string block redundant avionics and selective cross strapping. The OFS is a 3-axis stabilized spacecraft. The OFS has two primary configurations: aerocapture and orbital science. Figures 6, 7 and 8 illustrate the different configurations. Although all mechanism mass was include in the mass lists, Figures 6, 7, and 8 do not include some mechanism detail such as the HGA two axis gimbal assembly and the instrument platform two axis gimbal assembly. Table 4 summarizes the aerocapture system mass, defined for this paper to include the aeroshell structure, TPS, and propellant required for bank angle control and post aerocapture orbit adjust to reach the initial science orbit. Table 5 summarizes the orbiter post aerocapture dry mass.

The aerocapture system is 59% of the total entry mass. This include the TPS and aeroshell structure (~50% mass fraction) and enough propellant to perform bank angle control of the aeroshell during aerocapture and perform orbit adjust maneuvers post-aerocapture to achieve the desired science orbit of (altitude) 4,000 x 488,000 km. Twenty-four 66N SCAT Bi-Prop engines are used to provide up to 7.5 deg/sec^2 acceleration for bank angle control. The bi-prop engines are positioned and balanced to provide spacecraft torques about the velocity vector (40 deg angle of attack) even with one jet failed.

Table 4. Aerocapture System Mass

Mass in kg	Flt	CBE	Cont	MEV
Aerocapture System	34	898.2	24%	1119.1
Dry Mass	34	736.4	30%	957.3
TPS: Nose	1	204.0	30%	265.1
TPS: Windward	1	292.5	30%	380.3
TPS: Leeside Nose	1	56.0	30%	72.8
TPS: Leeside & Base	1	58.1	30%	75.5
TPS Adhesive	1	6.3	30%	8.2
Upper Structure	1	42.8	30%	55.6
Lower Structure	1	44.9	30%	58.4
Base Structure	1	7.2	30%	9.3
Separation Springs & H/W	26	24.7	30%	32.1
Propellant Mass		161.8		161.8
Bank Angle Control		22.5		22.5
4000x430000 dV (m/s)	360	115.9		115.9
4000x488000 dV (m/s)	78	23.4		23.4

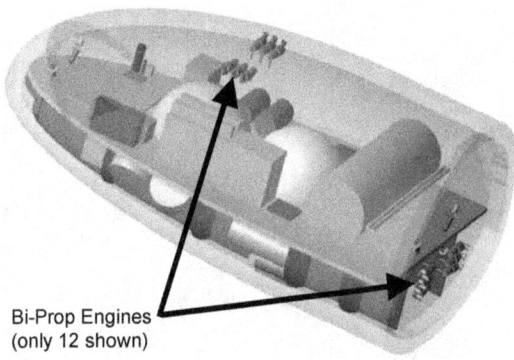

Bi-Prop Engines (only 12 shown)

Figure 6. Aerocapture Configuration

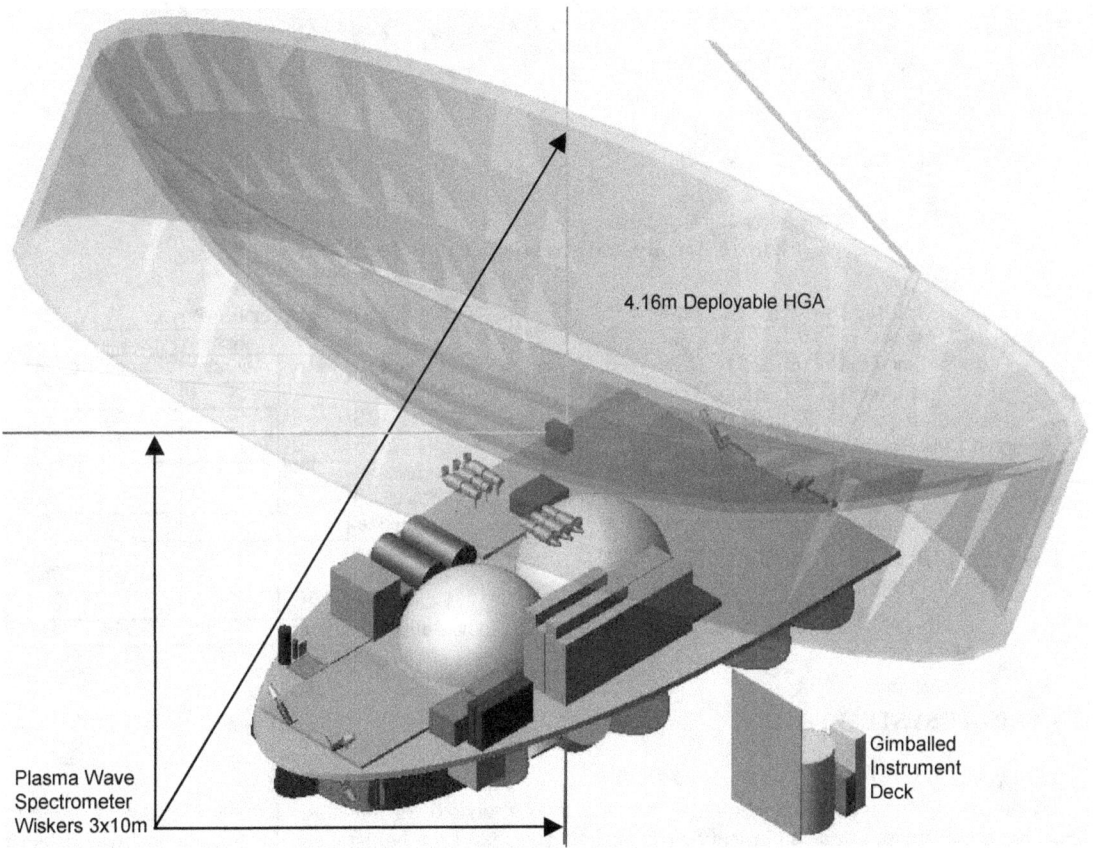

4.16m Deployable HGA

Gimballed Instrument Deck

Plasma Wave Spectrometer Wiskers 3x10m

Figure 7. Orbit Configuration Isometric

Table 6 summarizes the orbiter power modes. The modes listed are not all the modes identified in the study, just the ones that stress the system. Heater power in all phases is minimal because of an aggressive assumption that the MMRTG excess heat, ~2600W, can be distributed across the spacecraft well enough to not require the heater power typical for deep space missions. The available power listed is the power output of two dual stage MMRTGs (14% efficiency) with 1.5% output degradation per year. It is assumed that once the telecom and instrument components have been turned on, that they are never turned completely off, but rather are placed in a low power standby mode when not in use.

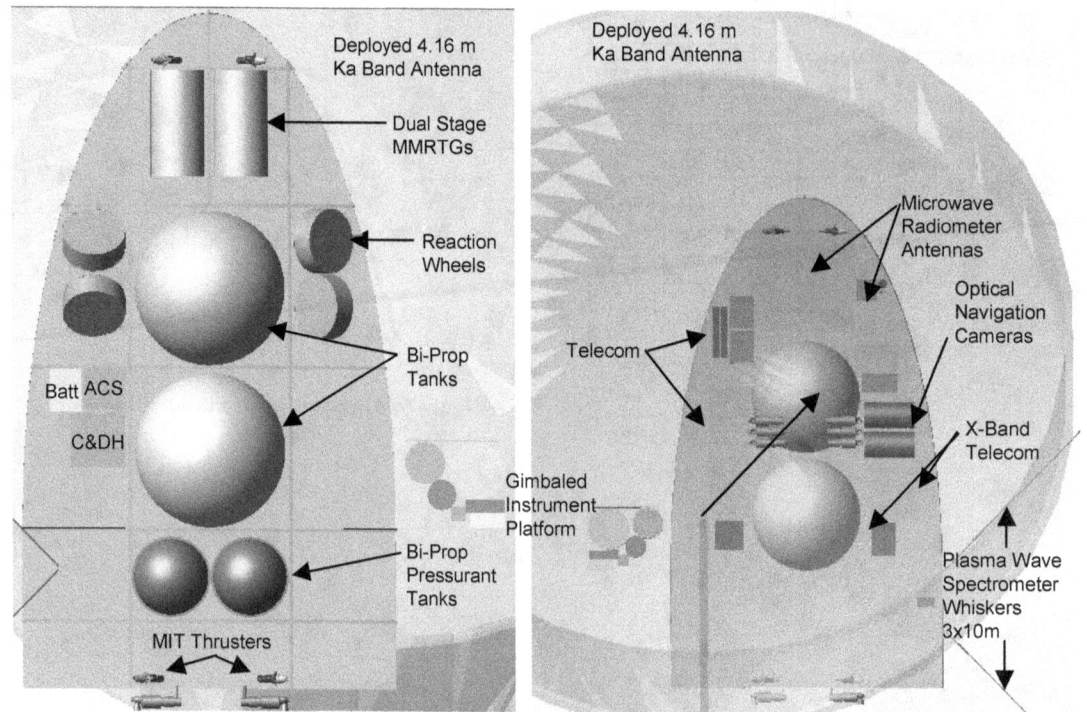

Figure 8. Orbit Configuration, Top/Bottom Views

Table 5. Orbiter Dry Mass Summary

Mass in kg	Flt	CBE	Cont	MEV
Orbiter Dry Mass	**236**	**524.2**	**27.3%**	**667.5**
Instruments	20	33.8	30.0%	43.9
Articulation & Att Control	28	53.7	12.0%	60.2
Command & Data Handling	32	34.7	27.0%	44.0
Power	4	64.7	30.0%	84.1
Telecom	23	41.9	28.5%	53.8
Structure	10	96.6	30.0%	125.5
Dual Mode Bi-Propulsion	118	104.9	27.5%	133.7
Thermal	1	54.0	30.0%	70.2
Harness	1	40.0	30.0%	52.0

Table 6. Orbiter Average Power (W)

Component	Mission Phase			
	Relay	Aerocap	Science	Comm
Instruments	0	0	30	24
ACS	88	69	78	78
C&DS	24	38	24	24
Telecom	90	13	13	38
Propulsion	0	90	50	50
Thermal	20	0	20	20
CBE Totals	222	210	214	234
MEV (20%)	**266**	**252**	**257**	**280**
Mission Year	10	10	12	12
Available Power (W)	357	357	344	344
Margin	**34.2%**	**41.9%**	**34.1%**	**22.8%**

KEY FLIGHT SYSTEM TRADES

AEROSHELL SHAPE AND SIZE

A discussion of the various aeroshell shapes and sizes analyzed for this study is extensive out of scope for this paper, but are discussed in detail in Ref 5. The desired aeroshell L/D, ballistic coefficient, and stability effect shape, size, and center of mass placement for the aeroshell; all of which affect volumetric and mass packaging efficiency of the internal components. For this study, a 0.8 L/D flattened ellipsled was selected.

Although the flattened ellipsled has the best volumetric efficiency (volume/surface area), the center of mass (CM) constraints associated with the desired angle of attack (40 deg, tends to want CM close to mid point between nose and tail) and stability (tends to want CM to be lower than the widest part of aeroshell) yield a large volume in the upper back region of the aeroshell that cannot be effectively utilized because of the need to offset any mass at the back of the aeroshell with mass in the nose. Although the 0.8 L/D flattened ellipsled works for this study, there may be opportunity to reduce overall vehicle mass by using some of the other alternative shapes examined.

The communications frequency (L-Band chosen), probe lifetime (total vertical descent as function of ballistic coefficient), and probe latitude / longitude placement (concurrent vs separate relay, antenna pointing error) all drive the relay antenna(s) design. The current L-Band design is a compromise between UHF (good visibility through atmosphere, but antenna too large) and X-Band (terrible visibility through atmosphere, but reasonable antenna size). Because the probes are not tracked after release, the communications link has to accommodate the expected pointing error over the duration of the probe lifetime (+/- 3 deg).

Deployable UHF or L-Band antennas could be used to increase useable surface area on the SEP stage and overall data capability.

ORBITER STRUCTURE

The primary structure of the orbiter had to accommodate launch modes and stresses, aerocapture stresses, and mechanical constraints associated with separating from the LV, the SEP stage, and aeroshell. It is possible that a series of smaller decks oriented in the horizontal plane with respect to the launch configuration might provide a better foundation for efficient spacecraft packaging.

SUBSYSTEM DESCRIPTIONS

Orbiter subsystems will be discussed in order of overall system impact. In general, subsystems discussed first drive the system design more than those discussed last.

AEROCAPTURE SYSTEM

The aerocapture system is defined as the TPS, the underlying aeroshell structure, and the propellant required to achieve the desired science orbit. The aerocapture system structure, TPS and their associated aero-thermal design basis are described in more detail in Ref 8-10. In summary, the aerothermal, aerodynamic, guidance, control analysis, and structural mass converged on a 2.88m flattened ellipsled with a L/D of 0.8.

The TPS was separated into 4 constant thickness zones to save mass: nose, windward side, forward leeside, and aft leeside (which includes the baseplate). For aeroshell jettison after aerocapture, the nose and windward zones were considered a single unit. The base plate would be jettisoned first, and the leeside and windward sides would be jettisoned concurrently.

24 bi-prop engines were used for bank angle control. This large number of engines allows discrete levels of torque to provide a range of small impulse corrections to large accelerations (7.5 deg/sec2). The engine configuration also provides redundancy for one engine out capability. These engines remain with the Orbiter after the aeroshell is jettisoned to eliminate the need for a separate aerocapture propulsion system.

Approximately 161 kg of propellant is required for bank angle control during aerocapture, periapsis raise maneuver at the first apoapsis after aerocapture, and an apoapsis raise maneuver to place the orbiter in an orbit properly phased for a Triton flyby resonance.

TELECOM

The Orbiter telecom system includes X, Ka, and L-Band components. The L-Band components are required for atmospheric probe relay; L-Band is capable of penetrating Neptune's atmosphere at the desired science altitudes. The X and Ka bands are used for Earth Communication, X for safe mode and Ka for science data return. Table 7 summarizes the driving data return links. The L-Band antenna is approximately 1.6 m x 1.6 m with a mass of about 8 kg. This antenna is attached to the SEP stage with a single axis gimbal to point to the probes during probe entry.

The Ka link uses a 35 W TWTA in combination with a deployable 4.16m HGA (59.6 dB) and is capable of returning ~270 Gbits of data assuming 4 hours of comm. per day to a 70m DSN station. This provides ample margin to the 151 Gbits of planned science data. X-Band link uses a 15 W SSPA with a small dish antenna (32.2 dB) to

Table 7. Orbiter Telecommunications Links

Probe to Orbiter Relay					Probe			Orbiter	
Mission Phase	Freq	Max Range	Data	Excess	Power	Ant	Off	Ant	Off
Probe 1 Relay	L-Band	255000 (km)	80	3	20	0	80	25.1	3
Probe 2 Relay	L-Band	133000 (km)	120	4	20	0	80	25.1	3
Orbiter to Earth					Orbiter			DSN	
Probe Relay	Ka-Band	31	400	3	35	41.6	0.12	70M	
Launch	X-Band	0.25	12.5	3	15	2	45	70M	
Cruise	X-Band	31	100	3	15	32.2	0.36	70M	
Orbit Science	Ka-Band	31	25600	3	35	59.6	0.01	70M	

provide Earth communication up to SEP stage separation. The L-Band probe relay link uses a 20W SSPA with a 25.1 dB antenna up to 324 Kbits total data return from the first probe and up to 648 Kbits total from the second probe.

POWER

Two dual stage Mutli-Mission Radioisotopic Thermal Generator (MMRTG) units, generating 3000W thermal, were selected for the Orbiter power source. The single stage MMRTG is planned for TRL 6 by 2006, laboratory tests currently have the dual stage technology at TRL 2-3 and funding profiles plan TRL 6 by 2010. The expected performance of the dual stage MMRTG is approximately 14% efficiency with 1.5% degradation per year providing 420W at beginning of life and 344W at end of life.

Secondary batteries are included to help during peak periods with a typical assortment of battery charge controllers, power switching, and power conversion electronics.

ACS

The Orbiter is a 3-axis controlled spacecraft that uses reaction wheels for attitude pointing and hydrazine thrusters for attitude maneuvers and reaction wheel de-saturation. The primary optical instrument deck is on a two axis gimbal as is the deployed HGA. Sun sensors are not utilized because of there no time critical sun point requirement (MMRTG power source). Star trackers are utilized in an orthogonal mount configuration to provide better attitude knowledge. Both the star trackers and the IMU have aggressive mass and powered consistent with low TRL units that are funded for achieving TRL 6 in the next decade. The current configuration could easily handle the mass of conventional units.

PROPULSION

The propulsion system is dual mode bi-prop system with a re-pressurization system. Hydrazine thrusters are used for short and infrequent pulse ACS duties; bi-prop engines are used for longer duration impulses such as aerocapture bank angle control and orbit adjust maneuvers. Thruster configurations are designed to accommodate a single thruster failure by removing an entire thruster string from the system.

C&DH

The C&DH system is based on a RAD750, 3U Compact PCI implementation that is currently offered in various forms by multiple vendors. Although some of the cards required may not yet exist in 3U format, it is assumed that they will by time of mission implementation, or can be developed with little difficulty.

THERMAL

The mission design presents several challenges for the thermal design:
1. The MMRTGs together generate ~2600W of waste heat.
2. The MMRTGs are enclosed in an aeroshell designed to keep heat from getting in (making it harder to get heat out).
3. The radiator system has to be designed to work before, during, and after aerocapture
4. Inside the aeroshell, the system will experience solar distance of 0.7 AU (Venus) to 31 AU.

Titanium and water loop heat pipes (LHP) running to hot radiators mounted on the SEP stage were chosen to solve the problem of getting the heat out of the aeroshell. Aluminum and ammonia loop heat pipes were also added to transport Orbiter electronics heat out of the aeroshell. A second set of titanium / water LHP carries MMRTG heat directly to the hydrazine tank.

From the previous year's Titan aerocapture study (Ref 12), a ~30 node lumped mass model of a spacecraft in an aeroshell using these LHP concepts was constructed to compute temperature distributions during the key mission phases for various design options. The computational model results confirmed that all of the key avionics and propulsion components were maintained well within prescribed operating temperatures during both the cruise to Saturn and after orbit insertion when the aeroshell was jettisoned and the orbiter exposed to the cold space environment at Titan. Although this analysis was not updated for the Neptune spacecraft, the results of the Titan study provide confidence that a similar design can do the job.

STRUCTURE

The structure is discussed in more detail in Ref 10. In general, the Orbiter primary structural design was driven by:
1. LV frequencies and loads.
2. Aerocapture loads.
3. Aeroshell separation planes.
4. Desired aeroshell center of mass.

NEW TECHNOLOGY DEVELOPMENT

Other than aerothermal analysis tools and TPS manufacturing techniques, there were no enabling technologies identified to implement the flight system. The In Space Propulsion Program is continuing funding to improve the aerothermal analysis tools. The dual stage MMRTGs, deployable Ka band HGA, and deployable square rigger solar arrays are strongly enhancing technologies, but these technologies are already independently funded for development and there are other options to implement the mission if these technologies do not become a reality.

RECOMMENDED ADDITIONAL ANALYSIS

Many questions and trades consistent with continued Phase A/B efforts were identified by the study team. A summary of these issues is presented below along with a general classification of the issue as a lien, or opportunity, or either.

- **Launch Vehicle**: There is so much unallocated margin on the Delta-IV Heavy that an Atlas V 551 could be feasible, especially if an Earth gravity assist is considered (opportunity).
- **Science Instruments**: 1) Develop conceptual designs for instruments and verify TRL, mass, power, volume estimates (Lien). 2) Verify optical, and radiative fields of view for all instrument (especially thermal radiative for instrument on gimbaled deck, and RF for microwave radiometer antennas).
- **Power**: 1) Develop detailed power modes and profiles (either). 2) Verify 2 MMRTGs are adequate for full mission (Lein). 3) Verify EMI/EMC compatibility for component configuration (either).
- **Thermal**: Verify MMRTG heat can be effectively routed to other spacecraft components to eliminate need for heaters (Lein).
- **Telecom**: 1) Add LGA/MGA for Earth acquisition prior to high bandwidth links (Lien).
- **Aeroshell**: 1) Re-investigate best shape for volumetric efficiency including center of mass location (opportunity). 2) Verify heating and TPS for new ballistic coefficient (either).
- **Cost**: Generate cost estimate for complete flight system (lien).

CONCLUSIONS

The study demonstrates general technical feasibility for a Neptune Orbiter flight system designed to use aerocapture as the orbit insertion mechanism. Many liens exist against the conceptual design presented, but opportunities and large launch mass margins balance the liens. Technology readiness for the flight system is good with all major components currently being funded to achieve TRL 6 in the next decade.

ACKNOWLEDGEMENTS

The work described in this paper was performed at the Jet Propulsion Laboratory, California Institute of Technology, under a contract to the National Aeronautics and Space Administration. This work was funded by the NASA/Marshall Space Flight Center (MSFC) In-Space Propulsion Program managed by Les Johnson (MSFC). Personnel (authors for Ref 1-10) from Langley Research Center, Johnson Space Center, Ames Research Center, Marshall Space Flight Center, and the Jet Propulsion Laboratory were instrumental in determining system level requirements and subsystem capabilities for trade and mission performance analysis. Discipline experts at the Jet Propulsion Laboratory including Dave Hansen (Telecom), John Huang (Antennas), Ray Baker (Propulsion), and Bill Nesmith (MMRTGs) all provided valuable input into the subsystem conceptual designs represented in this paper.

REFERENCES

[1]Lockwood, M.K., "Neptune Aerocapture Systems Analysis," AIAA-2004-4951.

[2]Noca, M. and Bailey, R. W., "Mission Trades from the Perspective of Aerocapture", AIAA-2004-3843.

[3]Haw, R. "Aerocapture Navigation at Neptune," AAS-03-643, August 3-7, 2003.

[4]Justus, C., Duval, A., Keller, V., "Atmospheric Models for Aerocapture Systems Studies," AIAA-2004-4952, August 16-19, 2004.

[5]Edquist, K., Hoffman, D., Rea, J., "Configuration, Aerodynamics, and Stability Analysis for a Neptune Aerocapture Orbiter," AIAA-2004-4953, August 16-19, 2004.

[6]Masciarelli, J., Westhelle, C., Graves, C., "Aerocapture Guidance Performance for the Neptune Orbiter," AIAA-2004-4954, August 16-19, 2004.

[7]Starr, B., Powell, R., "Aerocapture Performance Analysis for a Neptune- Triton Exploration Mission," AIAA-2004-4955, August 16-19, 2004.

[8]Hollis, B., Olejniczak, J., Wright, M., Takashima, N., Sutton, K., Prabhu, D., "Preliminary Convective- Radiative Heating Environments for a Neptune Aerocapture Mission," AIAA- 2004-5177, August 16-19, 2004.

[9]Laub, B., Chen, Y., "TPS Challenges for Neptune Aerocapture," AIAA-2004-5178, August 16-19, 2004.

[10]Dyke, R., Hrinda, G., "Structural Design for a Neptune Aerocapture Mission," AIAA-2004-5179, August 16-19, 2004.

[11]Lockwood, M.K., "Titan Aerocapture Systems Analysis," AIAA-2003-4799, July, 2003.

[12]Bailey, R.W., Hall, J.L., and Spilker, T.R, "Titan Aerocapture Mission and Spacecraft Design Overview," AIAA-2003-4800, July 22, 2003.

MISSION TRADES FOR AEROCAPTURE AT NEPTUNE

Muriel A. Noca and Robert W. Bailey

Jet Propulsion Laboratory, California Institute of Technology, Pasadena, California 91109, USA

A detailed Neptune aerocapture systems analysis and spacecraft design study was performed to improve our understanding of the techonology requirement for such a hard mission. The primary objective was to engineer a point design based on blunt body aeroshell technology and quantitatively assess feasibility and performance. This paper reviews the launch vehicle, propulsion, and trajectory options to reach Neptune in the 2015-2020 time frame using aerocapture and all-propulsive vehicles. It establishes the range of entry conditions that would be consistent with delivering a ~ 1900 kg total entry vehicle maximum expected mass to Neptune including a ~ 790 kg orbiter maximum expected mass to the science orbit. Two Neptune probes would be also be delivered prior to the aerocapture maneuver. Results show that inertial entry velocities in the range of 28 to 30 km/s are to be expected for chemical and solar electric propulsion options with several gravity assists (combinations of Venus, Earth and Jupiter gravity assists). Trip times range from approximately 10-11 years for aerocapture orbiters to 15 years for all-propulsive vehicles. This paper shows that the use of aerocapture enables this mission given the payload to deliver around Neptune compared to an all-propulsive orbit insertion approach. However, an all-propulsive chemical insertion option is possible for lower payload masses than the one needed for this science mission. Both approaches require a Delta IV heavy class launch vehicle.

INTRODUCTION

As part of the NASA In-Space Propulsion Program, aerocapture was investigated as an option for orbit insertion around Neptune. This study involved several NASA centers and had for objective to conceptually design an aerocapture system for a generic orbiter with atmospheric probes mission. This paper provides an overview of the mission trades performed during this study. The main objectives of the mission trades were to:

1) Identify potential mission architecture and trajectories for a launch circa 2015-2020, which meant to identify launch vehicle options, launch opportunities and sensitivities, and potential trajectories using chemical ballistic propulsion and solar electric propulsion (SEP);
2) Understand the sensitivities in flight time and Neptune atmosphere's inertial entry velocities;
3) Provide a baseline trajectory.

The level of analysis for the mission trades varied from relatively detailed, in the case of the aerocapture system and trajectory optimization, to more parametric in the case of the chemical system design. The approach was to survey as much as possible the trajectory trade space, both for chemical with multiple gravity assists and for SEP with a wide range of flight times and various gravity assist options. Once the trajectories were compiled, the delivered mass at Neptune was calculated given the maximum performances of representative launch vehicles. This delivered mass was then compared to the actual mass needed for an aerocapture vehicle and for a chemical insertion vehicle, thus quantifying the benefits of aerocapture.

This paper first summarizes the transportation architectures considered. It then describes the launch and transit options for ballistic chemical trajectories and SEP trajectories. These trajectories provide the range of inertial entry velocities to expect. Both aerocapture and chemical insertion are then discussed, including details about the systems and trajectories. Based on the systems and orbiter designs, Earth-Neptune trajectories can be picked and traded against. This paper finally shows the overall architecture trade results.

These trades do not represent all possible options for mission design or mission architecture, but they do span the range of likely options.

TRANSPORTATION ARCHITECTURES

The science objectives and basic spacecraft concept for this Neptune mission were based on previous studies performed internally at the Jet Propulsion Laboratory.[1] The mission includes two Neptune atmospheric probes and a Neptune Orbiter.[2] The atmospheric probes were considered here as black boxes (207 kg maximum expected value total), and only the navigation aspects of carrying these probes were taken into account.[3] The probes perform a direct entry a few hours before the Orbiter's insertion. The desired science orbit around Neptune is a 4,000 x 488,000 km altitude at 157 degree inclination to match Triton's orbit plane and motion (Triton's orbit is almost circular at 330,000 km radius from Neptune). The science orbit provides a fly-by of Triton every 3 orbits (~12 days).

To understand the sensitivities in aerocapture entry conditions into Neptune's atmosphere and understand the benefits of aerocapture, it was necessary to perform a trade study of the various and most probable transportation options to Neptune in the 2015-20 launch time frame. The transportation architectures considered from Earth to the final science orbit were the following:

1. **Option 1A**: Chemical ballistic transit trajectory to Neptune with a chemical insertion at Neptune.

2. **Option 1B**: Chemical ballistic transit trajectory to Neptune with aerocapture.

3. **Option 2A**: SEP transit trajectory to Neptune with a chemical insertion at Neptune.

4. **Option 2B**: SEP transit trajectory to Neptune with aerocapture.

The Orbiter was designed to perform aerocapture and modified in the trades when a chemical insertion was performed instead. As will be discussed, the baseline concept uses SEP to reach Neptune. To make the comparison between aerocapture and chemical insertion easier, and as the orbiter was designed to accommodate the ΔV for apoapsis raise, the targeted apoapsis altitude for the chemical insertion was 430,000 km (consistent with the target aerocapture altitude). The aerocapture entry is assumed to start at 1000 km altitude (25,766 km radius). The aerocapture pass is retrograde to match Triton's orbit. The chemical insertion burn is assumed to be performed at 4000 km altitude (28,766 km radius).

All four combinations of transportation were evaluated and will be described. The technique of aerobraking was not included in this trade study because the need for a highly elliptical science orbit precludes the multi-pass orbit circularization strategy that makes aerobraking useful. The next two sections will describe the transit trajectories, followed by a description of the orbit insertions.

EARTH TO NEPTUNE CHEMICAL BALLISTIC TRAJECTORIES

Ballistic direct trajectories as well as gravity assist trajectories were computed for a launch period between 2012 and 2019. The gravity assists (GA) include several combinations of Venus (V), Earth (E), Jupiter (J) and Saturn (S) fly-bys. Figure 1 shows the launch C3 for each of the trajectories surveyed. This list of trajectories does not represent all possible options of gravity assist nor launch date, and a more thorough survey should be done to complement this analysis.

As can be seen in Figure 1, the direct trajectories require a very high launch C3, which implies a relatively small launch mass. The Jupiter gravity assist (JGA) and Jupiter Saturn gravity assist also require a large launch C3 and thus offer poor performances. For all practical purposes, these trajectory options were deleted from the trade space.

Note that some of gravity assist trajectories require a deep space maneuver, which ΔV can be significant. Figure 2 shows the corresponding deep space maneuver ΔV required for each of the trajectories.

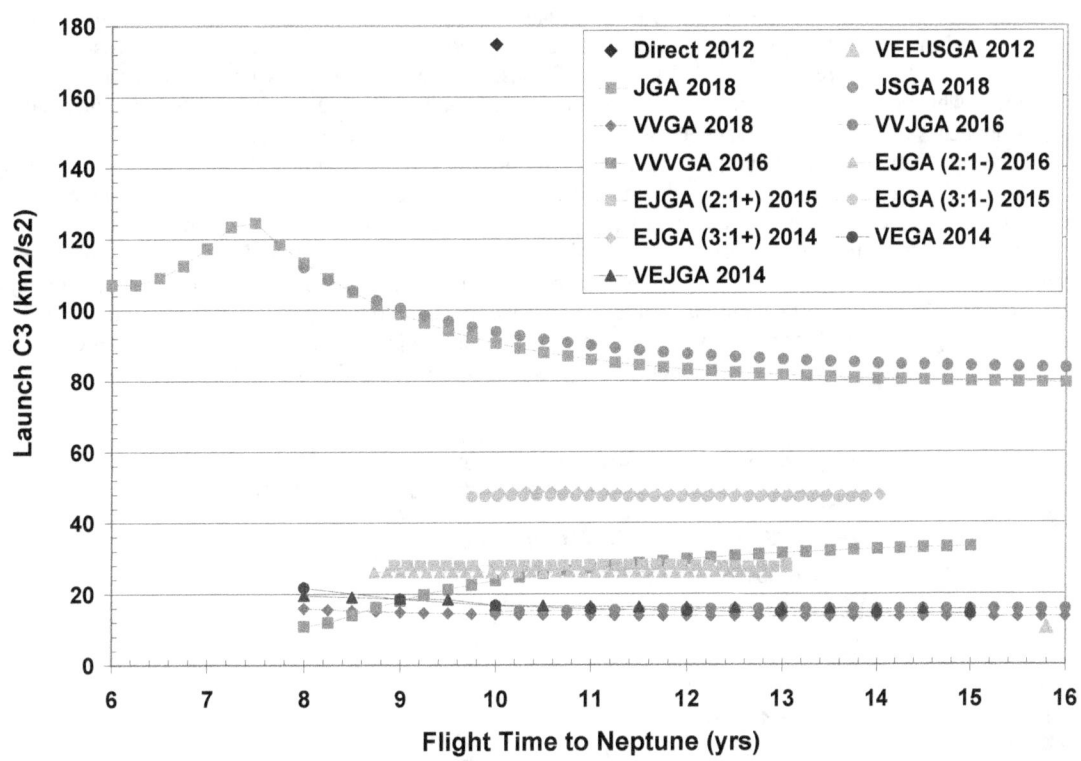

Figure 1: Launch C3 for various direct and GA trajectories to Neptune (with associated launch dates).

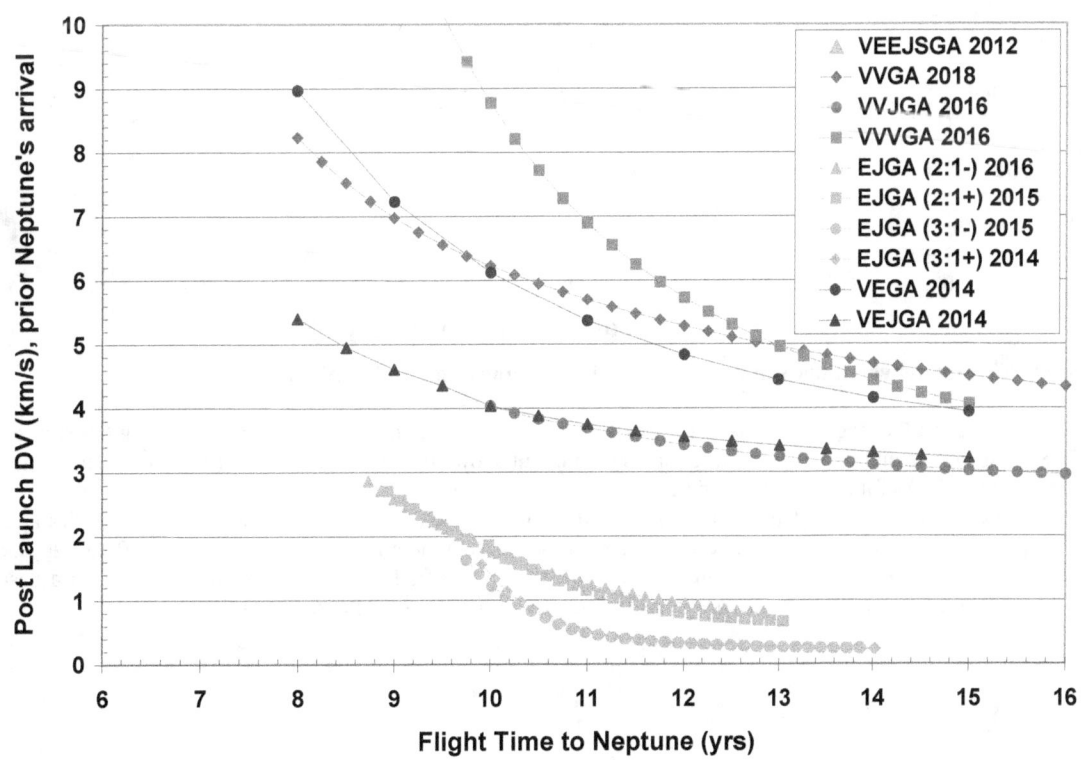

Figure 2: Corresponding post-launch ΔV for the gravity assist trajectories to Neptune.

Although the Venus only gravity assists offer a low launch C3, the post launch ΔV required for these options is quite high. The best performing options are the Venus or Earth combined with a Jupiter gravity assist. The launch date availability for these is restrained to 2014 – 2016. The use of a Jupiter gravity assist requires proper positioning of the planets, which only happens every ~ 10 years for Jupiter. Again, gravity assist with different launch dates and lower post launch ΔV may exist with launch C3 close to or lower than the ones presented here.

The maximum launch injected mass was derived from Figure 1 given the launch C3 for a Delta 4450, Atlas 551 and a Delta IV Heavy. These launch vehicles were picked as representative of a range of launch vehicle performance. The launch vehicle data was provided by the NASA KSC Launch Support Group[4]. Out of the injected mass provided by the launch vehicle, one needs to subtract the mass of propellant and chemical propulsion system needed to perform the post-launch ΔV. To do that we assumed a chemical Isp of 325 sec and a propulsion system dry mass equal to 20% of the propellant mass (approximately equivalent to a propulsion dry mass fraction of 16% of the chemical stage). Figure 3 then shows the net delivered mass (injected mass minus chemical propellant and propulsion dry mass to perform the post-launch ΔV) at Neptune's arrival (not inserted into Neptune's orbit).

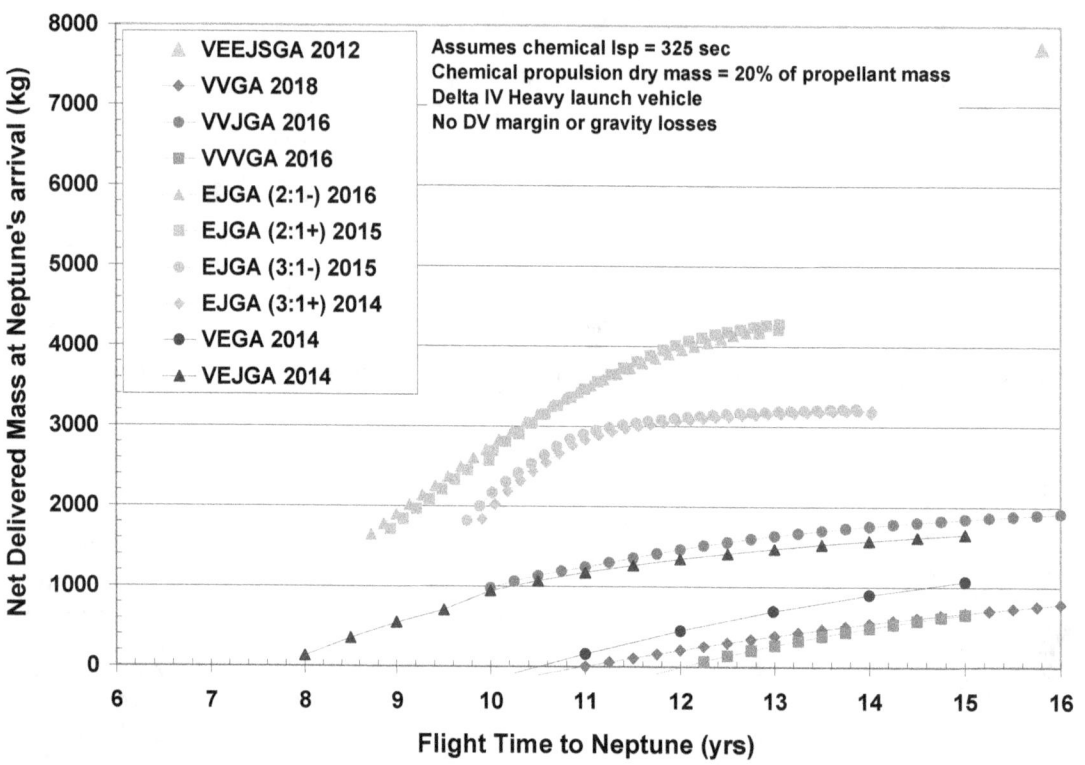

Figure 3: Net delivered mass at Neptune's arrival. Launch on a Delta IV Heavy.

Figure 3 shows that the EJGAs have the best potential for delivering significant masses around Neptune. The VEEJSGA is unique and very dependent on launch date, but shows that there could be trajectories performing even better than the EJGAs for a longer flight time.

The corresponding inertial entry velocities at a 1000 km altitude are provided in Figure 4. With both Figures 3 and 4, and assuming the EJGAs provide sufficient mass at Neptune's arrival, flight times of 10 to 12 years to Neptune imply inertial entry velocities in the range of 28 – 31 km/s. This information will help select a baseline inertial entry velocity for a detailed aerocapture design, as will be discussed later.

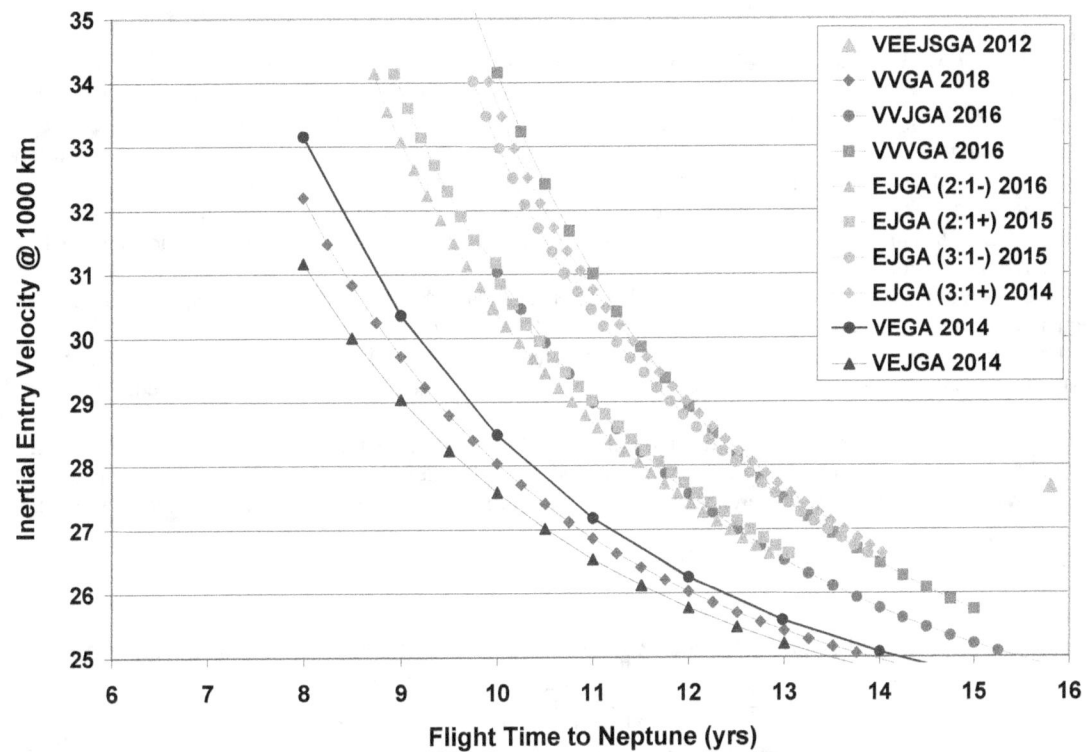

Figure 4: Inertial entry velocity for the ballistic gravity assist trajectories.

EARTH TO NEPTUNE SOLAR ELECTRIC PROPULSION TRAJECTORIES

As for the chemical ballistic trajectories, an extensive database of gravity assists Solar Electric Propulsion (SEP) trajectories on various launch vehicles was built, as they clearly provided better delivered mass for equivalent flight times compared to direct (no gravity assist) SEP trajectories. These trajectories served the purpose of evaluating the sensitivities in launch date, number of thrusters, power levels and inertial entry velocities.

The SEP low-thrust trajectory optimization were run with a code named SEPTOP for Solar Electric Propulsion Trajectory Optimization Program, which is based on the calculus of variations. This code optimizes two body interplanetary trajectories and can model discrete numbers of operating Xenon thrusters throughout the trajectory. The trajectories allowed for a coast time duty cycle of 5% to simulate times when the spacecraft is not thrusting due to housekeeping activities, and assumed a constant 250 W from the solar arrays for the spacecraft. A 10% launch vehicle margin was assumed.

SOLAR ELECTRIC PROPULSION SYSTEM ASSUMPTIONS

The ion thruster used to calculate the SEP trajectories is the NEXT engine. The characteristics of the NEXT engine technology can be found in many references.[5,6] Table 1 shows the projected performances of the NEXT engine. The high-Isp profile of the engine was used to calculate the trajectories.

The ion propulsion system (IPS) was designed more as a propulsion module than just thrusters and power processing units. Figure 5 shows a simplified block diagram of a typical single string ion propulsion system (IPS). To that basic configuration was added redundancy, structural and thermal considerations. Figure 5 also shows an example of what the IPS module designed here could look like.

The number of thrusters and PPUs was calculated on the basis of power requirements and thruster propellant throughput. The system architecture followed a conventional approach with parallel strings of PPUs and thrusters. Each PPU drives one thruster but is cross-strapped to two engines. One spare ion engine, one spare PPU and DCIU were also included for single-fault tolerance. Each thruster was gimbaled separately. The PPUs were assumed to be 95% efficient.

The solar arrays were sized based on a projection of the AEC-Able Square Rigger array capability. Since this array technology scales with power from ~ 1 kW up to ~ 30 kW, it was used as a representative potential technology for SEP applications. The specific mass was assumed to be 130 W/kg. A 14% degradation factor was applied to the array Beginning-of-Life (BOL) power to account for various degradation phenomena. Also, in order to support power demand during launch, a primary battery was used prior to solar array deployment.

The tank mass fraction was assumed to be 2.5% for Xenon when stored as a supercritical gas (~2000 psia). Furthermore, a 10% propellant contingency was added to the deterministic propellant mass to account for flow rate characterization, residuals, attitude control and margin.

Since the system masses are function of mainly power level, launch mass and propellant mass, each trajectory was uniquely considered and had a system mass associated with it. The component and subsystem sizing assumptions are given in Table 2. To be consistent with the JPL Team X conceptual design guidelines at the time of the study, 30% mass contingency was applied to all spacecraft subsystems. These masses represent the Maximum Expected Value mass (MEV).

For the purpose of the trade studies, the mass model of Table 2 was used. However, the final baseline design and the final table comparing all options (section VII) included a better definition of the system. This system was about 35 kg heavier than the model used here. Most of the difference resides in a larger thermal and structural mass.

Table 1: High-level NEXT thruster characteristics as compared to the flown DS1 ion engine NSTAR.

	NSTAR DS1	NEXT
Max. thruster processed power (kW)	0.5 - 2.3	1.2 - 6.2
Engine diameter (cm)	30	40
Maximum Isp (sec)	3100	3900
Xe throughput per engine (kg)	130	250
Thruster mass (kg)	8.3	12
Power Processing Unit mass (kg)	11.9	21

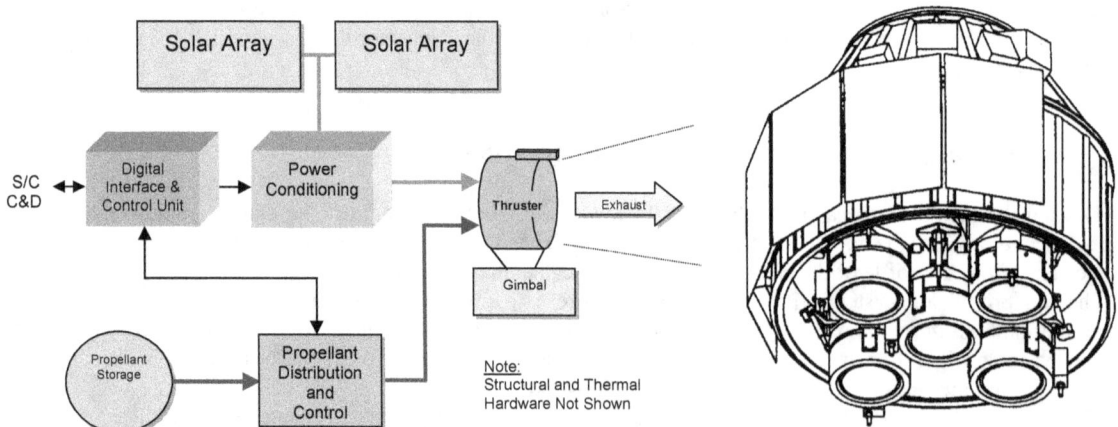

Figure 5: Ion propulsion module block diagram and conceptual configuration for system sizing

Table 2: Example of a 30-kW ion propulsion system mass model (includes 30% contingency).

Subsystem	Maximum Expected Mass (kg)
Not scaled with propellant mass: - Propulsion	263
- Power	415
- Cabling	90
- Thermal	69
- Telecom+ACS +electronics	102
- Structure	390
Scaled with propellant mass: - Tank	2.5%
- Tank structure	4%
- Thermal	1%

SEP TRAJECTORY PERFORMANCES AND INERTIAL ENTRY VELOCITY CHARACTERIZATION

As previously mentioned, the SEP trajectories considered use one to several gravity assist, because the mass needed for the orbiter and orbiter insertion system was beyond what a direct SEP trajectory could provide. With the appropriate thruster model, trajectories were run for several power levels, launch vehicle types and launch years. Results are in terms of net delivered mass. The net delivered mass is defined as the spacecraft mass minus the dry mass of the ion propulsion system and minus the propellant mass used for the SEP transfer. Therefore the net delivered mass is everything on the spacecraft that isn't SEP propellant or part of the ion propulsion module. Each trajectories were optimized for maximum delivered dry mass at destination (the launch C3 and thrusting profile were optimized). The gravity assist opportunities were also optimized for a given launch year. Figure 6 shows results of net delivered mass as a function of flight time to Neptune (arrival at Neptune's vicinity, not the final science orbit around Neptune), launch vehicle, power level and launch years for the NEXT thruster system.

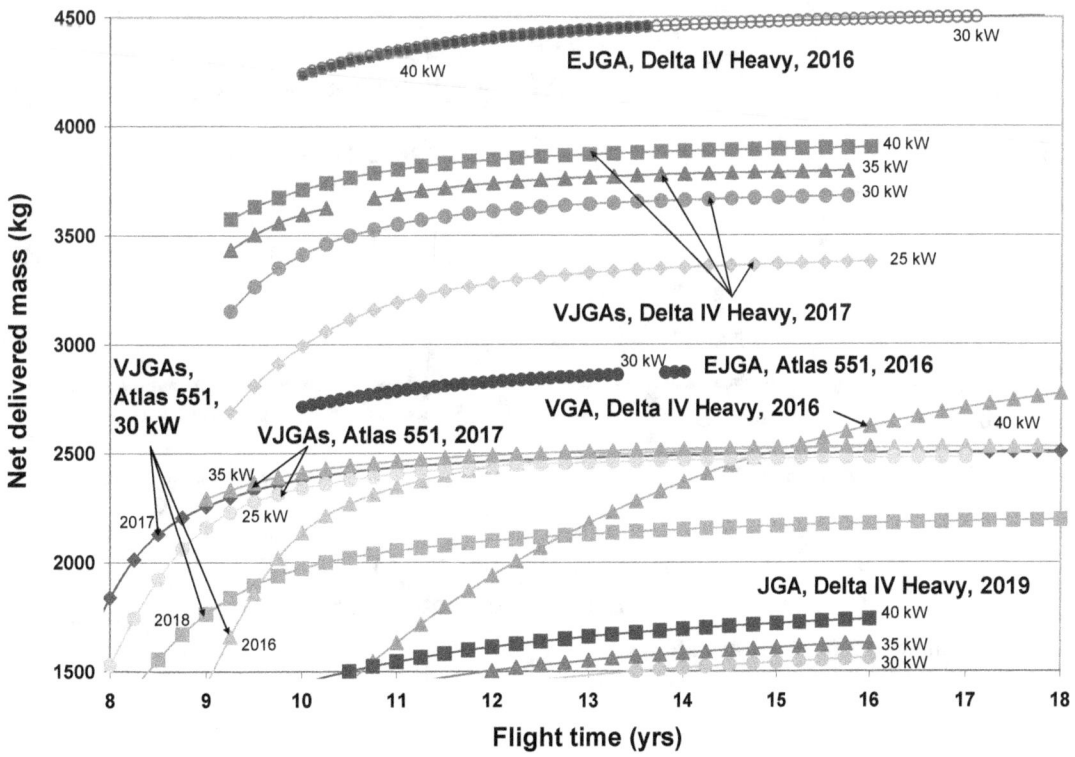

Figure 6: Net delivered mass at Neptune's arrival. SEP trajectories.

35

As can be seen in Figure 6, the net delivered mass is a strong function of launch year, gravity assist combination and power level. However, it is believed that the cases shown here are quite representative of the range of performance that can be expected from an SEP system. Not all of the trajectories run are shown in this figure as some provided performances below 1900 kg, and thus were judged insufficient.

Figure 7, 8 and 9 show the corresponding launch C3, radius of Jupiter fly-by and inertial entry velocities at 1000 km altitude at Neptune, respectively. The radius of the Jupiter fly-by was unconstrained and thus optimized in this study. For the VJGAs and EJGAs of particular interest, the fly-by occurs typically between the orbits of Io and Europa, so significant radiation can be expected.

As Figure 9 shows, the inertial entry velocity increases is very dependent on flight time, launch date, and gravity assist. However, the inertial entry velocity is only weakly dependent on SEP power for a given launch date and thruster technology. Thus choosing a flight time range will determine a range of inertial entry velocities. Flight times between 10 to 12 years offer the most "net delivered mass" benefit and result in entry velocities less than 30 km/s for most launch opportunities.

The weak sensitivity to the SEP power level in inertial entry velocity is mostly due to the fact that over the range of power looked at, the trajectory optimization code is trying to follow the same optimum acceleration path. Thus for high power level, it will optimize the trajectory at lower launch C3, thus injecting more mass. The acceleration, which is proportional to the power level to mass ratio will be roughly the same as a low power, large C3, low launch mass case. Since it will follow almost the same trajectory profile, the arrival hyperbolic velocity will only vary slightly. This is the case for a fixed flight time and launch date.

The launch window to perform a given gravity assist is about one month. The sensitivity in propellant mass for that window is included in the 10% propellant margin.

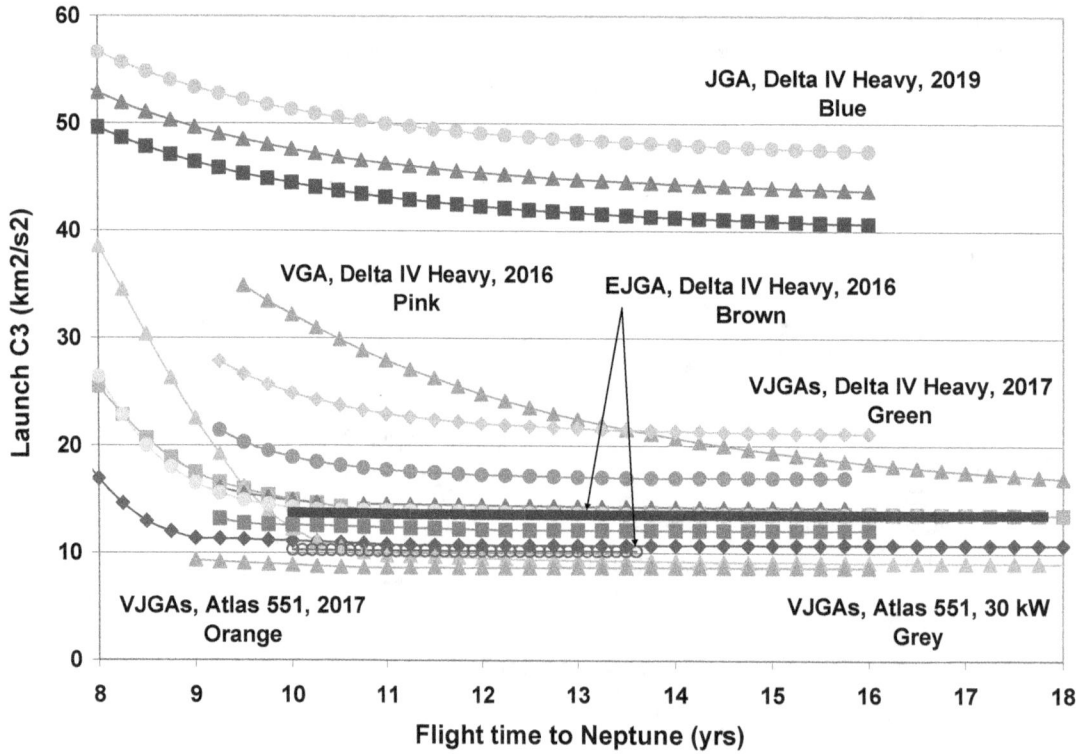

Figure 7: Launch C3 for the SEP trajectories.

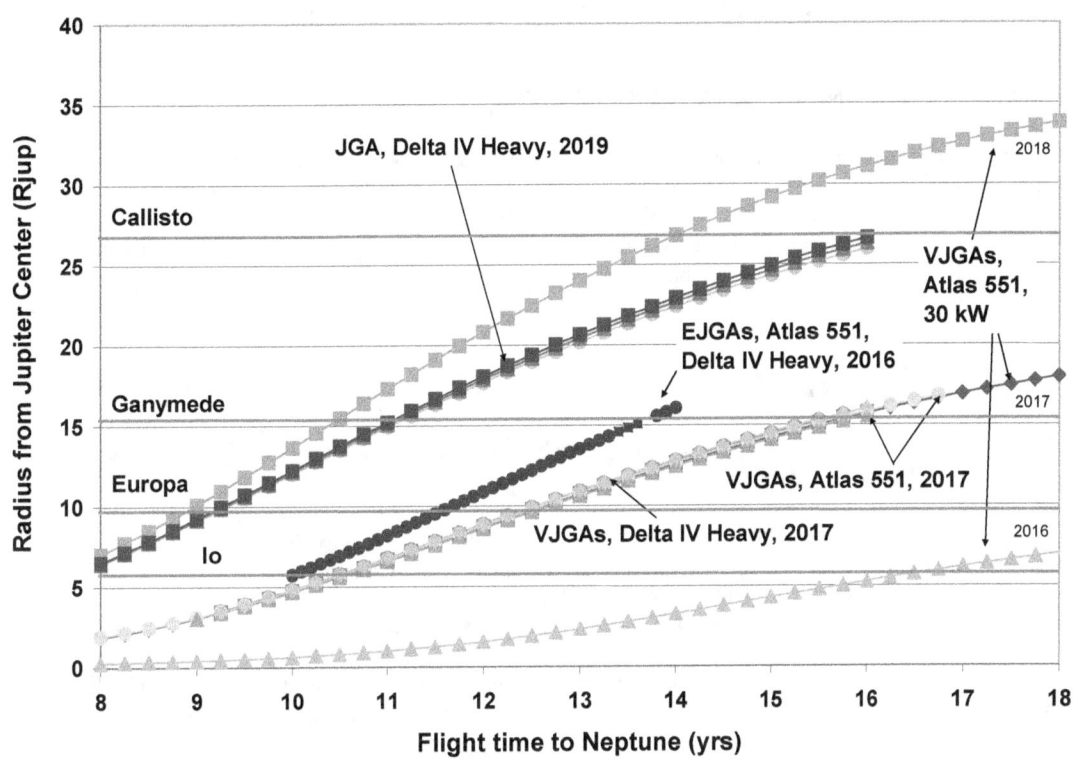

Figure 8: Jupiter gravity assist fly-by distance for the SEP trajectories.

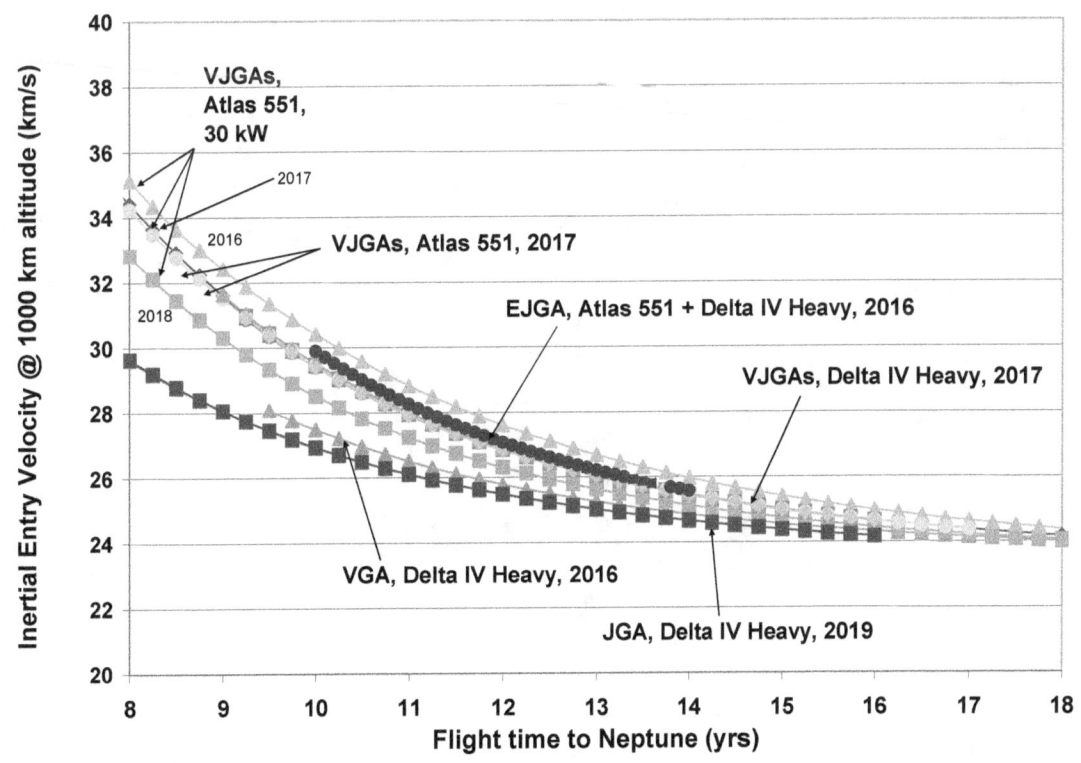

Figure 9: Inertial entry velocity for the SEP trajectories.

AEROCAPTURE SYSTEM AND INSERTION (OPTIONS 1B AND 2B)

SELECTION OF THE INERTIAL ENTRY VELOCITY

The selection of the chemical ballistic or SEP trajectory Earth to Neptune transit trajectory depends mostly on the performance (net delivered mass) provided by the trajectory and is an iterative process since the mass desired at Neptune's arrival depends on the design of the aerocapture system. In both chemical ballistic and SEP cases, an entry velocity of 29 km/s represents the best compromise between short flight times and high net delivered masses. Although somewhat arbitrary, it was felt that the aerocapture design would not change significantly for inertial entry velocities between 28 and 30 km/s.

BASELINE TRAJECTORY

The baseline trajectory for the design of the aeroshell and other components of the aerocapture system was selected based on the following criteria:

1. The mission architecture should use the smallest launch vehicle possible to reduce cost;
2. The trajectory performance should provide adequate system mass margin (30%) for maximum expected mass, and adequate system reserves (> 10%);
3. The trajectory should provide a Neptune inertial entry velocity close to 29 km/s.

At the time of selection, the SEP trajectories were providing better performance than chemical ballistic option. It was also felt that a Venus Jupiter gravity assist would be sufficient performance wise. Thus the trajectory selected featured (see Figure 10):

Launch vehicle:	Delta IV Heavy (5 m fairing)
Flight time:	10.25 years
Launch date:	2/17/2017
Arrival date:	5/20/2027
Gravity Assist:	Venus Jupiter
Launch C3:	18.4 km^2/s^2
Launch mass:	5964 kg
Propellant mass:	973 kg deterministic
Vhyp @ Neptune:	17.5 km/s
$V_{entry\,inertial}$:	29 km/s @ 1000 km
Thrusters:	5 maximum operating NEXT
SEP power level:	30 kW (1 AU EOL)

10.25 Year VJGA Neptune Flyby
Delta IV (4050H-19) / SEP 30 kW

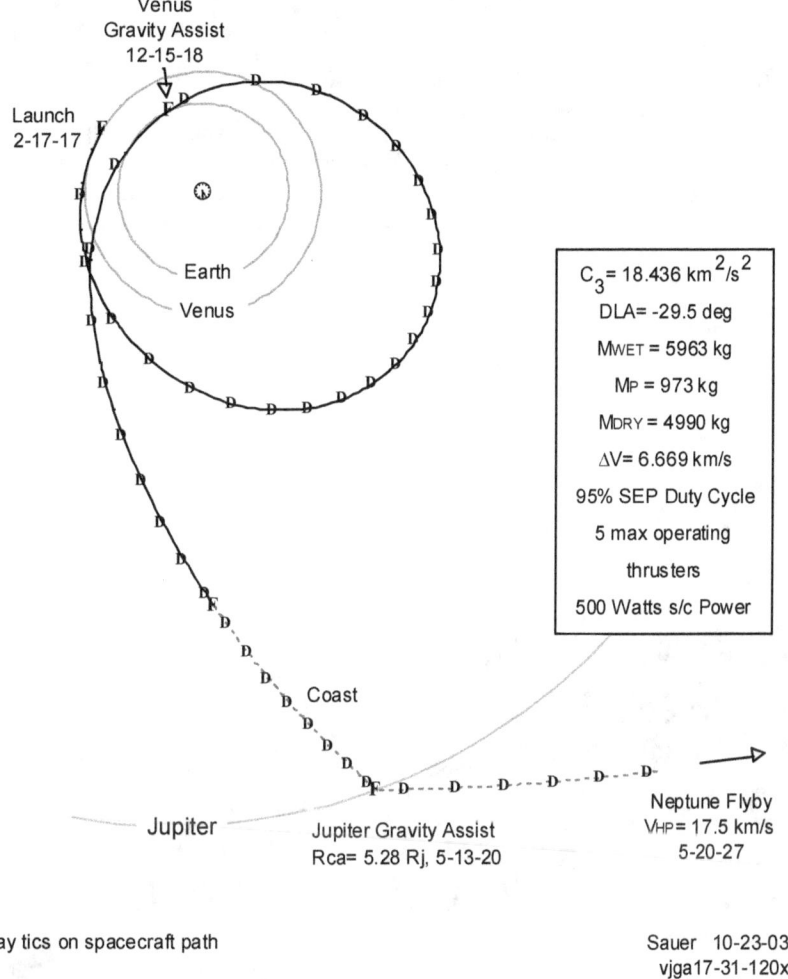

C_3 = 18.436 km^2/s^2

DLA = -29.5 deg

M_{WET} = 5963 kg

M_P = 973 kg

M_{DRY} = 4990 kg

ΔV = 6.669 km/s

95% SEP Duty Cycle

5 max operating

thrusters

500 Watts s/c Power

30 day tics on spacecraft path

Sauer   10-23-03
vjga17-31-120x

Figure 10: SEP baseline trajectory to Neptune.

AEROCAPTURE SYSTEM AND NEPTUNE ORBITER DESCRIPTION

The aerocapture system and the orbiter are described in detail in reference [2] and [7]. They have been designed for the baseline trajectory, which was chosen to match an inertial entry velocity of 29 km/s. The aerocapture system and the aeroshell shape were subject to an intensive trade related to the aerothermal and aerodynamic properties, control authority issues and volumetric efficiency. The design converged on a 2.88 m ellipsled with a L/D of 0.8. Figures 11 and 12 show the SEP cruise configuration and the post-SEP and aerocapture configuration respectively.

The thermal protection system was optimized to reduce mass. The aeroshell is jettisoned after aerocapture, and a dual mode bi-propellant system was used to raise the pariapsis. A ΔV of 438 m/s was used to size the propellant tank and loading. This ΔV includes the maneuvers to be performed during the aerocapture pass, a periapsis raise maneuver at first apoapsis and and apoapsis raise maneuver to place the orbiter in phase for Triton fly-bys.

Table 3 summarizes the mass breakdown for the aerocapture system As can be seen, the total dry mass of the aeroshell system is 1119 kg (maximum expected mass) for a total entry maximum expected dry mass of 1911 kg (~ 58.6% aerocapture system entry dry mass fraction). Other aerocapture-related hardware was ejected before entry. This hardware is also summarized in Table 3.

39

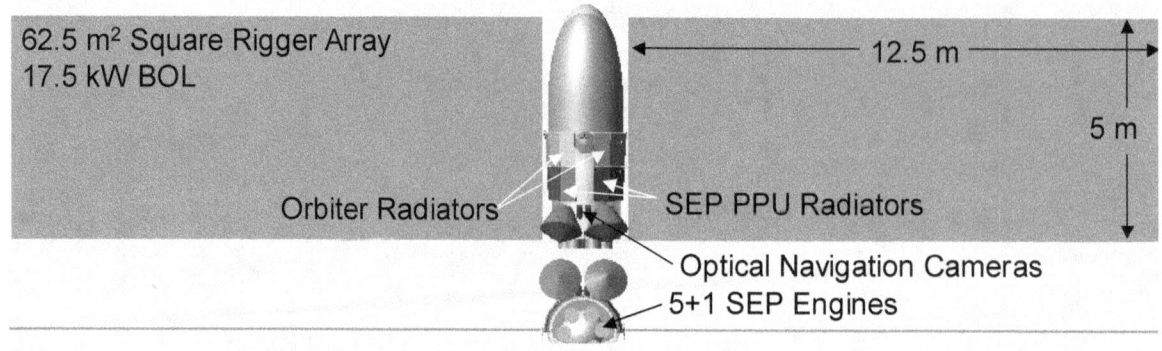

Figure 11: Neptune orbiter SEP cruise configuration.

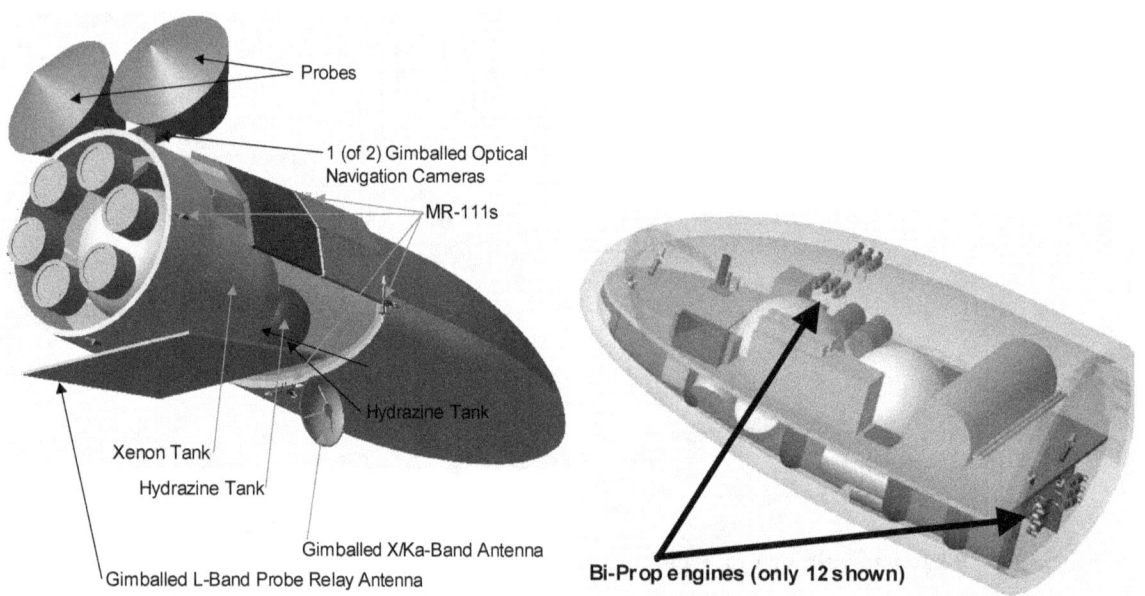

Figure 12: Neptune orbiter post-SEP and aerocapture configuration.

Table 3: Aerocapture system mass breakdown (includes 30% contingency).

Subsystem	Maximum Expected Mass (kg)
Mass that entered the atmosphere:	
- Heatshield, backshell and structure	957
- Hydrazine propellant	162
Aerocapture mass jettisoned prior entry:	
- 2 entry probes and ~ 100 kg of hydrazine	207 + 111 = 318

CHEMICAL SYSTEM AND INSERTION (OPTIONS 1A AND 2A)

SELECTION OF THE TRANSIT TRAJECTORY AND CHEMICAL INSERTION

The selection of the chemical ballistic or SEP trajectory Earth to Neptune transit trajectory depends on the performance (net delivered mass) provided by the trajectory but also very strongly on the Neptune Orbit Insertion (NOI) burn. Unlike aerocapture, the chemical insertion will be very dependent on the arrival velocity (Vinfinity). For the chemical ballistic transfer trajectories, a balance between the post-launch ΔV and the NOI ΔV needs to be taken into account to minimize the sum of both ΔVs.

Given the chemical ballistic gravity assist trajectories provided in Figures 1 and 2 and their arrival velocity at Neptune, one can calculate the NOI ΔV at 4000 km altitude. Figure 13 provides such ΔV. A similar plot could be done for the SEP transit trajectories.

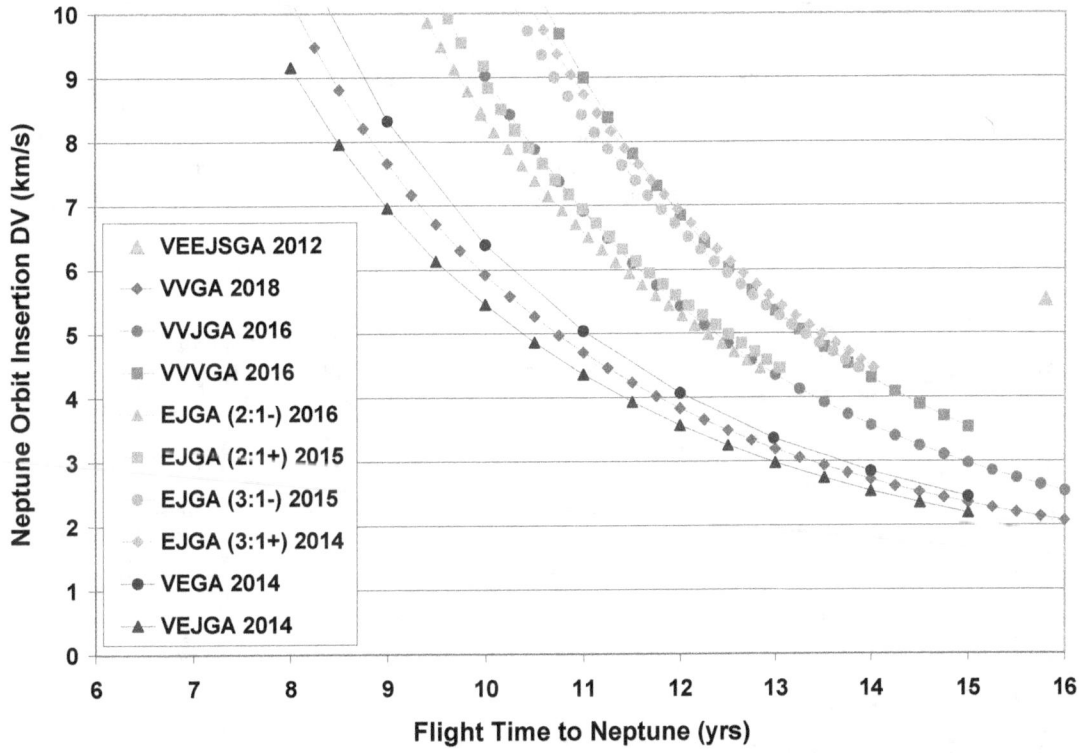

Figure 13: Neptune orbit insertion ΔV for the chemical ballistic transfer trajectories.

Since the NOI ΔV are large (which motivates the development of aerocapture), we assumed a dedicated chemical stage for that maneuver. In the case of chemical transit trajectories, we combined both ΔVs in a single stage (might provide conservative results).

CHEMICAL PROPULSION SYSTEM

To perform the chemical deep space maneuvers or insertion burns, a generic bi-propellant system was assumed, staged as necessary to accommodate for the large ΔVs. The dry mass for this system is summarized in Table 4. The specific impulse of the chemical system was assumed at 325 sec. In addition, 5% of the deterministic propellant mass was held as propellant contingency for maneuver clean-ups. Some of the structural mass depends on the dry mass at the beginning of the ΔV phase (Mi).

Table 4: Chemical propulsion system mass breakdown (includes 30% contingency).

Subsystem	Maximum Expected Mass (kg)
Not scaled with propellant mass:	
- Propulsion	19.5
- Thermal	16.5
- Telecom+electronics	2.3
- Structure	71.5 + 4% of initial mass (Mi)
Scaled with propellant mass:	
- Tank	5%
- Tank structure	4%
- Thermal	1%

Given these assumptions, the net delivered mass in a 4000 x 430,000 km altitude science orbit could be calculated. Figure 14 shows the net delivered mass for the various chemical ballistic trajectories. This figure represents option 1A, an all chemical propulsion to the science orbit. The "best" case for an all propulsive option is a Venus-Earth-Jupiter gravity assist that can deliver around 500 kg in 15 years. The Earth-Jupiter gravity assists may offer better performances but no data was generated that would confirm the slope for longer flight times.

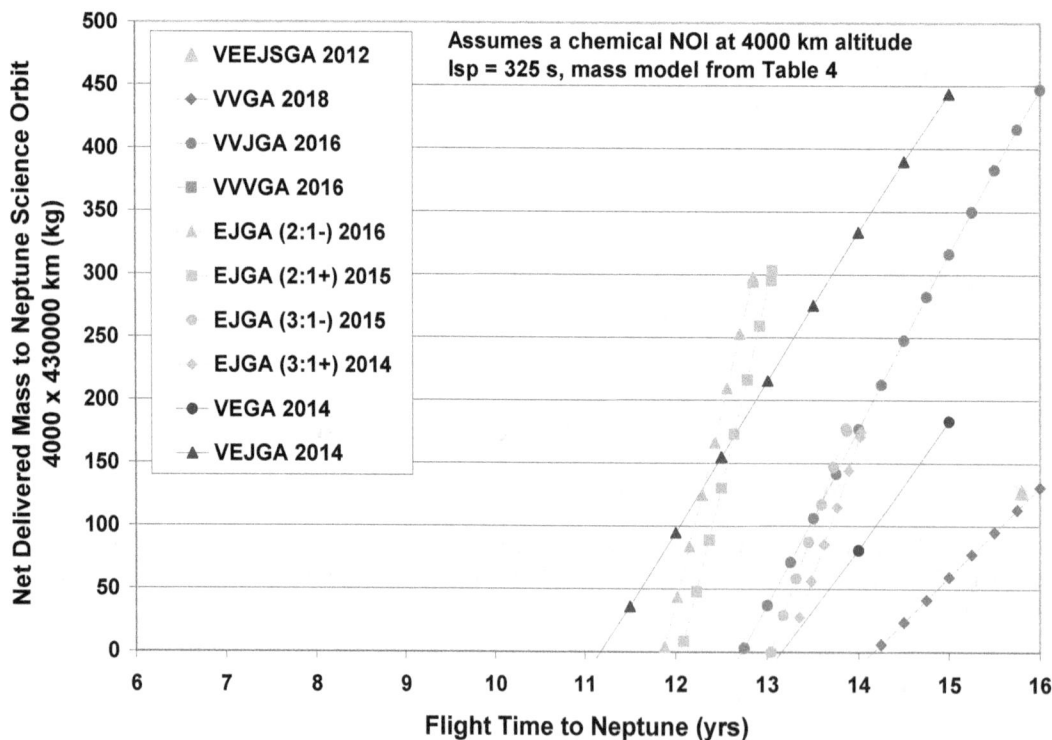

Figure 14: Net delivered mass in Neptune science orbit. Option A1 all chemical.

MISSION ARCHITECTURE TRADE RESULTS

The overall mission architecture trade results are summarized in Table 5. This table shows first the type of launch vehicle followed by the gravity assist type, the transit propulsion system and the Neptune capture system. It assumes that the full capability of the launch vehicle is used and calculates the payload surplus or deficit mass compared to the mass required at Saturn before insertion. The points picked in this table represent cases for each architecture option with an inertial entry velocity of 29 km/s, thus reducing the number of trajectory options.

In the case of chemical insertion, the orbiter to lander interface is assumed not to be jettisoned and thus is included in the payload in Neptune's orbit mass. The detailed mass breakdown can be found in [2].

The payload surplus or deficit mass is the mass above or below the necessary mass to deliver the atmospheric probes and orbiter around Neptune (system margin needs to be above 15-20%). Table 5 clearly show the advantages of aerocapture, which in every case looked at provided more payload reserve and shorter flight times than for a chemical insertion burn. However, they also show that it is possible to deliver sufficient payload mass (low margin) with a chemical insertion system. Here again, the penalty will be flight time.

Table 5: Architecture trades summary table.

Launch Vehicle	Delta IV H							Atlas 551	
Gravity Assist	VEJGA	EJGA				VJGA		EJGA	
Earth to Neptune Prop System	Chem	Chem		SEP		SEP		Chem	SEP
NOI Prop System	Chem	Aero	Aero	Chem	Aero	Chem	Aero	Aero	Aero
Option	A1	A2	A2	B1	B2	B1	B2	A2	B2
Cruise Time to Neptune (yrs)	15.0	10.8	11.8	15.0	10.5	15.0	10.3	11.8	10.5
Launch Year	2014	2016	2014	2016	2016	2017	2017	2014	2016
Launch C3 (km2/sec2)	15.6	26.0	47.3	13.5	13.6	17.0	18.4	47.3	9.1
SEP Power (kW, EOL)				30	30	30	30		30
Inertial Entry Velocity (km/s)		29	29		29		29	29	29
Neptune Cruise Chem DV (km/s)[1]	3429	1413	357					357	
NOI Chem DV (km/s)[1]	2300			2871		2781			
Launch Capability	7012	5695	3550	6543	6532	6130	5964	2630	4850
Propellant Mass[2,3]	4158	2040	376	655	809	1025	1070	279	713
LV to Prop Module Adapter	62	62	62	62	62	62	62	62	62
Prop Module Dry Mass	806	542	289	1437	1449	1465	1468	243	1441
Chem Prop Mod to Payload Adapter	40	40	40					40	
Pre-NOI Separated Mass[10]	318	318	318	318	318	318	318	318	318
Pre-NOI Net Delivered Mass	1628	2694	2464	4071	3895	3260	3046	1688	2315
Aerocapture System[4]		1119	1119		1119		1119	1119	1119
NOI Chem Propellant Mass[8]	966			2417		1898			
NOI Chem Dry Mass	280			487		413			
Payload in Neptune Orbit	792	792	792	792	792	792	792	792	792
System Margin = LV-MEV	(409)	783	553	375	1984	157	1135	(223)	404
System Margin % = (LV-MEV)/MEV	-5.5%	15.9%	18.5%	6.1%	43.6%	2.6%	23.5%	-7.8%	9.1%

MEV: Maximum Expected Value = best estimate + 30% contingency

Assumptions and Notes:

All masses are MEV mass listed in kg

[1] Includes 5% DV contingency

[2] Chem Propellant mass calculated using "Launch Capability" as system total mass; Chem Isp = 325 sec

[3] SEP Propellant mass calculated using "Launch Capability" as system total mass; includes 10% prop mass contingency

[4] Aerocapture System Mass: aeroshell structure, TPS, and DV to achieve 28766x488,000 km orbit

[6] Propellant mass and Prop Module Dry Mass for SEP / Chem options includes propellant and dry mass for both SEP and chemical stages

[7] Neptune Aerocapture Study Reference Mission

[8] Chem Propellant mass calculated using "Pre-NOI Net Delivered Mass" as Initial mass; Chem Isp = 325

[9] Total Cruise+NOI DV split equally between two stages; I.e. Cruise delta-V is staged

[10] Includes Probes and ~100kg of cruise hydrazine

CONCLUSIONS

This paper summarizes the transit trajectory options for a Neptune orbiter mission and derives the range of entry conditions for the aerocapture maneuver inside Neptune's atmosphere. This survey shows that inertial entry velocities in the range of 28 - 30 km/s are to be expected. This range offers the best combination of highest delivered mass to Neptune's orbit and lowest entry heating. The study chose to baseline an inertial entry velocity of 29 km/s for the detailed design of the aerocapture system, and the corresponding SEP trajectory is provided.

This paper also summarizes the mission transportation trades performed during the study to show the benefits of aerocapture. The study shows that aerocapture as an orbit insertion option provides more delivered mass in every launch vehicle and gravity assist case looked at than chemical insertion, and shorter flight time (typically by 4-5 years). However, all chemical or SEP with chemical insertion cases exist that would deliver about 450 kg in Neptune's orbit with a Delta IV Heavy with flight times around 15 years. The baseline trajectory case for this study is an SEP aerocapture case on a Delta IV heavy with a flight time of 10.2 years delivering a 790 kg orbiter.

ACKNOWLEDGMENTS

The research described in this paper was carried out at the Jet Propulsion Laboratory, California Institute of Technology, under a contract with the National Aeronautics and Space Administration.

The authors gratefully acknowledge the contributions of Carl Sauer, Jon Sims, and Theresa Debban-Kowalkowski for all their trajectory creativity and calculations. The authors also wish to thank Mary-Kae Lockwood (LaRC), Jeff Hall (JPL) for their conscientious review of the results, Steve Oleson (GRC) for all the SEP thruster details, Ray Baker (JPL) for all the chemical system details, and the In-Space Propulsion Program for making the funding available for this work.

REFERENCES

[1] T. Sweetser et al., "Titan Orbiter Team X report, June 4-15, 2001.

[2] R. Bailey, J. Hall, T. Spilker,N. Okongo, "Neptune Aerocapture Mission and Spacecraft Design Overview", AIAA-2004-3842, 40th AIAA/ASME/SAE/ASEE Joint Propulsion Conference, Fort Lauderdale, FL, July 2004.

[3] R. J. Haw, "Aerocapture Navigation at Neptune", AAS 03-643, AAS/AIAA Astrodynamics Specialists Conference, Big Sky Resort, Big Sky, Montana, August 3-7, 2003.

[4] NASA-KSC. Launch vehicle database. http://elvperf.ksc.nasa.gov/elvMap/ November 2002.

[5] S. Oleson, L. Gefert, S. Benson, M. Patterson, M. Noca, J. Sims, "Mission Advantages of NEXT: NASA's Evolutionary Xenon Thruster", AIAA-2002-3969, 38th AIAA/ASME/SAE/ASEE Joint Propulsion Conference, Indianapolis, IN, July 2002.

[6] M. J. Patterson, J. E. Foster, T. W. Haag, V. K. Rawlin, G. C. Soulas, R. F. Roman, "NEXT: NASA's Evolutionary Xenon Thruster", AIAA-2002-3832, 38th AIAA/ASME/SAE/ASEE Joint Propulsion Conference, Indianapolis, IN, July 2002.

[7] M. Lockwood, "Titan Aerocapture Systems Analysis", AIAA-2003-4799, 39th AIAA/ASME/ SAE/ASEE Joint Propulsion Conference, Huntsville, AL, July 2003.

CONFIGURATION, AERODYNAMICS, AND STABILITY ANALYSIS FOR A NEPTUNE AEROCAPTURE ORBITER

Karl T. Edquist

NASA Langley Research Center, Hampton, Virginia, 23681

Ramadas K. Prabhu

Lockheed-Martin Engineering and Sciences Company, Hampton, Virginia, 23681

David A. Hoffman and Jeremy R. Rea

NASA Johnson Space Center, Houston, Texas, 77058

A multi-center NASA team conducted a systems analysis study of a Neptune aerocapture orbiter mission in order to demonstrate feasibility and identify technology gaps. The aerocapture maneuver utilizes aerodynamic drag to decelerate the vehicle, rather than chemical propulsion, for orbit insertion around Neptune and allows a flyby of the Triton moon. This paper presents the analysis used to select an orbiter shape, and the aerodynamics and stability characteristics of the reference vehicle. Several shape classes were screened for aerodynamic performance using modified Newtonian theory. A lift-to-drag ratio requirement of 0.6 to 0.8 was derived from an estimate of the theoretical corridor width to give margin beyond 3-σ dispersions. A flat-bottomed ellipsled was selected as the reference orbiter shape based on various metrics, including lift-to-drag ratio, ballistic coefficient, and effective volume. High-fidelity computational solutions for the reference orbiter shape predict a lift-to-drag ratio of 0.806 and ballistic coefficient of 895 kg/m^2 at a trim angle-of-attack of 40 deg. Stable pitch behavior is predicted with a 6.2% static margin for an axial center of gravity at 51% of the vehicle length from the nose. Both the longitudinal short-period and lateral Dutch-roll frequencies are shown to be within acceptable limits based on piloted vehicle specifications. Aerodynamics uncertainties were estimated to result in a lift-to-drag ratio uncertainty of +13.4%/-14.5% using RSS values and +26.1%/-22.2% using stacked worst-case values.

NOMENCLATURE

C_A	=	axial force coefficient, $A / q_\infty S_{ref}$
C_D	=	drag coefficient, $D / q_\infty S_{ref}$
C_L	=	lift coefficient, $L / q_\infty S_{ref}$
C_l	=	rolling moment coefficient, $M_l / q_\infty S_{ref} L_{ref}$
C_{l_β}	=	derivative of rolling moment coefficient with respect to yaw, $\partial C_l / \partial \beta$ (rad^{-1})
C_m	=	pitching moment coefficient, $M_m / q_\infty S_{ref} L_{ref}$
C_{m_α}	=	derivative of pitching moment coefficient with respect to angle-of-attack, $\partial C_m / \partial \alpha$ (rad^{-1})
C_N	=	normal force coefficient, $N / q_\infty S_{ref}$
C_n	=	yawing moment coefficient, $M_n / q_\infty S_{ref} L_{ref}$
C_{n_β}	=	derivative of yawing moment coefficient with respect to yaw, $\partial C_n / \partial \beta$ (rad^{-1})
$C_{n_{\beta\text{-}dynamic}}$	=	lateral-directional stability parameter (rad^{-1})
C_p	=	pressure coefficient, $(p - p_\infty) / q_\infty$
C_Y	=	side force coefficient, $F_Y / q_\infty S_{ref}$
CG	=	center of gravity
h	=	altitude above 1 *bar* pressure level (km)
h_{stat}	=	static margin (% of L_{ref})
I_{xx}, I_{yy}, I_{zz}	=	moments of inertia $(kg\text{-}m^2)$
I_{xy}, I_{xz}, I_{yz}	=	products of inertia $(kg\text{-}m^2)$

L	=	aeroshell length (m)
L/D	=	lift-to-drag ratio
m	=	aerocapture mass (kg)
q	=	dynamic pressure, $\frac{1}{2}\rho V^2$ (Pa)
S	=	surface area (m^2)
T	=	temperature (K)
U	=	aeroshell volume (m^3)
V	=	velocity relative to atmosphere (km/s)
V_{eff}	=	effective volume
X	=	axial coordinate (m)
Y, Z	=	lateral coordinate (m)
α	=	trim angle of attack (deg)
β	=	yaw angle (deg)
β_m	=	ballistic coefficient, $m / C_D S_{ref}$ (kg/m^2)
γ	=	flight path angle (deg)
Ω	=	Dutch-roll frequency (rad/s)
ω	=	short-period frequency (rad/s)
θ	=	angle between surface normal vector and freestream velocity vector (deg)
ρ	=	density (kg/m^3)
σ	=	standard deviation

SUBSCRIPTS

cg	=	center of gravity
$base$	=	aeroshell base
$neut$	=	neutral point
ref	=	reference value
wet	=	wetted surface
∞	=	freestream condition

INTRODUCTION

A multi-center NASA team conducted a systems analysis study of a Neptune aerocapture orbiter mission[1-8] under funding from the In-Space Propulsion program. The study was conducted to demonstrate the benefits and feasibility of aerocapture, and identify technology gaps and requirements that must be addressed. Detailed analyses were performed in the areas of mission design, configuration, aerodynamics, propulsion, mass properties, structural dynamics, aerothermodynamics, thermal protection, atmospheric modeling, trajectory simulations, guidance, and navigation. This paper covers the trade study that resulted in the reference orbiter shape, as well as the static aerodynamics and a preliminary stability analysis of the vehicle.

Figure 1 shows the reference orbiter in its aerocapture configuration. The orbiter is designed to release two probes before atmospheric entry. The science orbit is designed to observe both Neptune and a nearby moon, Triton. An elliptical orbit with an apoapsis of 430,000 km was selected to achieve the science goals. *Aerocapture* is proposed as an alternative to chemical propulsion to insert the aeroshell into an elliptical orbit around Neptune. Aerocapture is a maneuver performed at the target planet in which the vehicle enters the atmosphere in order to remove kinetic energy via aerodynamic drag. The maneuver must be performed so that the vehicle reaches deep enough into the atmosphere to remove sufficient energy to allow capture and shallow enough to avoid hitting the planet. A small amount of chemical propulsion is needed to raise the periapsis and place the spacecraft into the final science orbit. Figure 2 shows a schematic of aerocapture at Neptune with a science orbit that intercepts Triton's orbit.

The aerocapture maneuver sets requirements on navigation accuracy in order to enter the atmosphere at an acceptable flight path angle (γ). The minimum γ results in the shallowest entry and the highest integrated aeroheating loads. The maximum γ gives the steepest atmospheric path, and results in the largest aeroheating rates and aerodynamic loads. The region between the minimum and maximum allowable γ determines the *entry corridor*. The orbiter lift-to-drag ratio (*L/D*) is chosen to provide adequate corridor width such that the vehicle can accommodate the 3-σ dispersions with margin. A preliminary analysis including uncertainty estimates for navigated entry states, atmospheric density, and aerodynamics showed that an *L/D* of 0.6 is just sufficient to capture the 3-σ entry corridor[1]. In order to have margin above 3-σ, an *L/D* of 0.8 was selected as the baseline requirement. The current analysis presents orbiter shapes with *L/D* = 0.6 - 0.8.

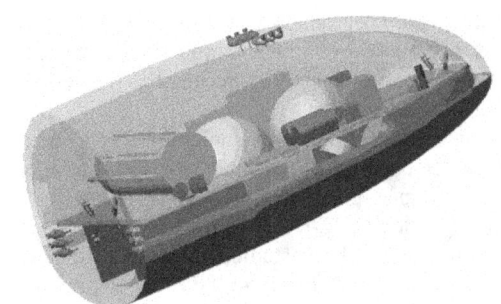

Figure 1. Neptune Orbiter in Aerocapture Configuration

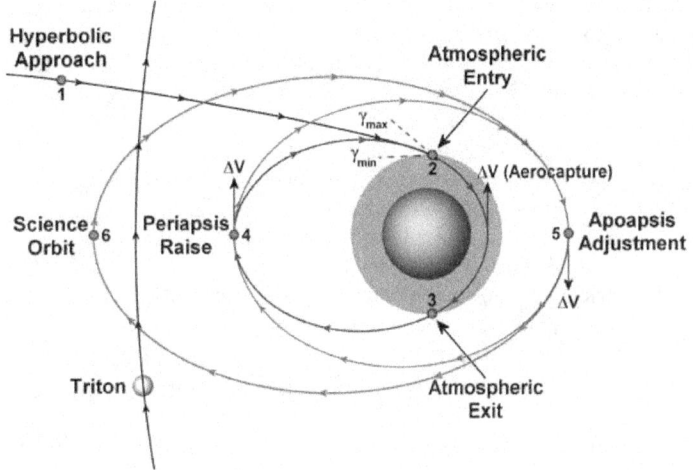

Figure 2. Schematic of Aerocapture Orbit Insertion at Neptune

ANALYSIS

ORBITER AEROSHELL SHAPE SELECTION

A *L/D* between 0.6 and 0.8 requires a vehicle shape that is more slender than typical 70-*deg* sphere-cone planetary entry vehicles, which give a maximum practical *L/D* near 0.25. The orbiter shape affects several other aspects of the mission, including aerodynamics, aeroheating, structures, packaging, mass properties, and thermal protection. Thus, considerable time was spent on an orbiter shape trade study before additional system analyses were performed. The shape trade study was undertaken in order to rapidly assess the performance capabilities of candidate mid-*L/D* shapes. The shapes were defined parametrically and aerodynamic performance was screened using *modified Newtonian Theory*, which is known to give reasonable predictions for blunt shapes at hypersonic speeds. In this fashion, many shapes were defined and analyzed in a short amount of time.

The goals of the shape study were to:

1. Achieve *L/D* = 0.6 - 0.8
2. Minimize *ballistic coefficient, β_m*
3. Maximize *effective volume*[9], *V_{eff}*

where *V_{eff}* is a measure of the effective internal packaging volume:

$$V_{eff} = \frac{6\sqrt{\pi}U}{S_{wet}^{3/2}} = 1 \text{ for a sphere} \qquad (1)$$

A *L/D* > 0.6 ensures that the orbiter can accommodate the 3-σ dispersions during aerocapture. Minimizing β_m reduces aeroheating rates and requirements placed on the entry guidance system. Maximizing *V_{eff}* gives the lowest surface area for a given volume, which can help reduce the aeroshell structure and thermal protection system (TPS) masses. Goal 1 tends to produce shapes that are long and slender, whereas goals 2 and 3 favor shorter, more compact shapes. Thus, a balance between these opposing characteristics is desired for the Neptune orbiter.

The candidate shape classes, examples of which are shown in Figure 3, were screened for the desired metrics. Variations on all shapes were performed parametrically. *Sphere-cones* have simple geometries and provide the *L/D* necessary for Neptune aerocapture. *Biconics* have previously been studied for aerocapture at Neptune and Mars, and have been shown to give good performance[10-11]. *Ellipsleds* that have been considered for a Mars lander application[12]. Bent biconics and modified ellipsleds were studied in order to increase *L/D* to 0.8.

Comparisons between the various shapes were performed for a fixed volume and mass. The allocated system entry mass was specified to be 2200 *kg*[1]. The shapes were screened using the following process:

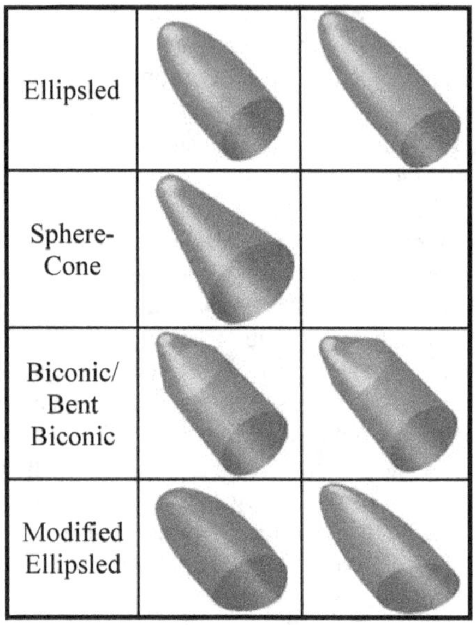

1. Fix the aeroshell mass and volume
2. Vary the geometric parameters depending on the shape class:
   a. Sphere-cone – total length, nose radius and cone angle
   b. Ellipsled – total length, nose length, and diameter
   c. Biconic – vary parameters manually
4. Calculate the aerodynamic coefficients using modified Newtonian Theory for $\alpha = 40 - 60$ *deg*
5. Identify shapes with the best combination of *L/D*, β_m, and V_{eff}
6. Estimate shape effects on aeroheating, structures, packaging, etc.

Ellipsled		
Sphere-Cone		
Biconic/ Bent Biconic		
Modified Ellipsled		

Figure 3. Examples of Shape Classes Considered for the Orbiter Aeroshell (*L/D* = 0.6 - 0.8)

Qualitative considerations were taken into account during the shape selection process. A small nose radius is undesirable from an aeroheating and packaging standpoint. A long slender shape is also not desirable because additional structural mass is needed to give sufficient stiffness for launch loads. Engineering judgment was used to account for these effects before additional high-fidelity analyses were performed.

MODIFIED NEWTONIAN THEORY

Modified Newtonian Theory was used in order to rapidly screen aerodynamic performance of the candidate orbiter shapes with reasonable accuracy. The theory produces good aerodynamics data for bodies at hypersonic speeds and is simple to implement. The theory expresses pressure coefficient (C_p) as a function of the angle between the local surface normal and the freestream velocity vector (θ):

$$C_p = C_{p_{max}} cos^2 \theta \quad (2)$$

where $C_{p_{max}}$ is evaluated behind a normal shock at the freestream Mach number. The effects of shear stresses are neglected, but they are small at hypersonic speeds. Given a discrete mesh of the surface distribution, aerodynamic coefficients can be estimated in a matter of seconds using a personal computer.

AEROCAPTURE DESIGN TRAJECTORY

After selection of the reference orbiter shape, high-fidelity computational fluid dynamics (CFD) solutions were used to predict detailed aerodynamic characteristics. Freestream conditions (density, velocity, and temperature) were needed from a design trajectory in order to run these solutions. For the current study, the *minimum density atmosphere*, *lift-up* trajectory with $\beta_m = 400$ *kg/m²* was selected for CFD analysis (Figure 4). The trajectory was calculated for an inertial entry velocity of 29 *km/s* and was obtained using the steepest entry flight path angle and the lowest atmospheric density expected based on the NeptuneGRAM model[2]. These conditions result in the highest convective heating rates experienced by the orbiter. Table 1 shows the freestream conditions used for the CFD solutions.

STATIC AERODYNAMICS

High-fidelity aerodynamics data were needed for the Monte-Carlo trajectory analysis and for a preliminary stability assessment of the reference orbiter. Viscous CFD solutions were obtained using the Langley Aerothermodynamic Upwind Relaxation Algorithm (LAURA)[13]. LAURA was developed at NASA Langley Research Center and has been used previously to predict the aerodynamic characteristics for various planetary[14] and space transportation[15] vehicles. The code uses a finite-volume approach to solve the viscous Navier-Stokes flowfield equations. A 3-species dissociating Neptune atmosphere model (H_2, H, He) was used to capture the high-temperature effects on hypersonic aerodynamics. The freestream composition was specified to be 68.2% H_2 and 31.8% He by mass. The entire aerocapture pass is hypersonic, so CFD solutions for a range of α at a high Mach number is considered sufficient for the current study. LAURA solutions at zero yaw angle were used in the longitudinal stability analysis.

Additional solutions at $\beta = 0$, 2, and 5 *deg* were run with the FELISA[16] CFD code for the lateral stability analysis. The FELISA software is a set of computer codes for the generation of unstructured grids around arbitrary bodies, simulation of three-dimensional steady inviscid flows, and post-processing of the results. The software has been extensively used for aerodynamic studies of the X-43, Mars landers, and similar configurations[17]. An equilibrium Neptune atmospheric gas was used in the present computations.

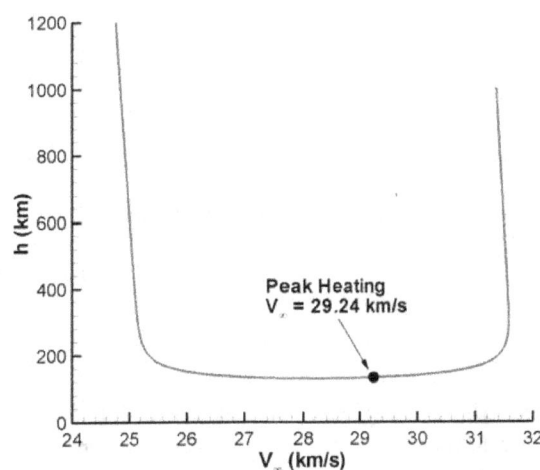

Figure 4. Aerocapture Entry Trajectory Altitude vs. Velocity

Table 1. Freestream Conditions for CFD Solutions

Parameter	Value
h (km)	132.2
V_∞ (km/s)	29.24
ρ_∞ (kg/m^3)	1.45 x 10^{-4}
T_∞ (K)	103.2

STABILITY ANALYSIS

The aerodynamic data for the Neptune orbiter were used in a simplified vehicle stability analysis. The analysis involved static stability, which determines whether a vehicle can fly in a trimmed equilibrium condition, and dynamic stability, which evaluates how a vehicle responds to aerodynamic disturbances. LAURA viscous data were used for the longitudinal static and dynamic stability analyses, and FELISA inviscid solutions at non-zero yaw were used for the lateral dynamic stability assessment.

The estimate of static longitudinal stability is determined by the pitching moment coefficient (C_m) about the vehicle's center of gravity (CG). If C_m for a particular flight condition (α and Mach number) is zero and its derivative with respect to α is negative ($C_{m_\alpha} < 0$), then the vehicle tends to trim at that α when flying at that speed. The degree of longitudinal static stability about a trim point is measured by the *static margin*, which can be defined in several ways. For this study, the static margin was taken to be the distance along the vehicle's longitudinal axis from the CG to the *neutral stability point* (X_{neut}), which is the axial CG location that gives $C_{m_\alpha} = 0$:

$$h_{stat} = 100 \times \frac{X_{neut} - X_{cg}}{L_{ref}} \qquad (3)$$

As long as the CG is forward of the neutral stability point, the vehicle will be statically stable at trim.

The dynamic stability analysis assesses the response of a statically stable vehicle flying in a trimmed condition to disturbing forces. For the simplified analysis presented here, the disturbances were assumed to be small and the vehicle's longitudinal and lateral–directional responses were assumed to be decoupled. Also, the damping terms were neglected at hypersonic Mach numbers. These assumptions reduced the analysis to the computation of two parameters: the short–period (ω) and the Dutch–roll (Ω) frequencies. The former is a measure of the vehicle's response to longitudinal disturbances and the latter provides an indication of its lateral–directional response to a disturbance. They are defined as:

$$\omega = \sqrt{\frac{-qS_{ref}L_{ref}C_{m_\alpha}}{I_{yy}}} \qquad (4)$$

$$\Omega = \sqrt{\frac{qS_{ref}L_{ref}C_{n_{\beta-dynamic}}}{I_{zz}}} \qquad (5)$$

where $C_{n_{\beta-dynamic}}$ is a measure of roll/yaw coupling and is positive for stable behavior:

$$C_{n_{\beta-dynamic}} = C_{n_\beta}\cos(\alpha) - \frac{I_{zz}}{I_{xx}}C_{l_\beta}\sin(\alpha) \qquad (6)$$

RESULTS AND DISCUSSION

ORBITER AEROSHELL SHAPE SELECTION

A reference orbiter shape was selected by parametrically screening candidate shapes using modified Newtonian aerodynamics and ranking the shapes according to the goals of $L/D = 0.6 - 0.8$, minimum β_m, and maximum V_{eff}. Figure 5 and Table 2 summarize the $L/D = 0.6$ shapes resulting from the study. In general, ellipsleds gave the best combination of the desired aerodynamic and geometric characteristics. Sphere-cones and biconics gave similar aerodynamic performance as the ellipsled, but they are generally longer and have less effective volume. The sphere-cone must fly at a lower α in order to achieve the same L/D as the ellipsled and biconic shapes. An ellipsled modified to have a flattened bottom is the shortest of all shapes, and has the lowest β_m and highest V_{eff}.

Table 2. Summary of $L/D = 0.6$ Shapes

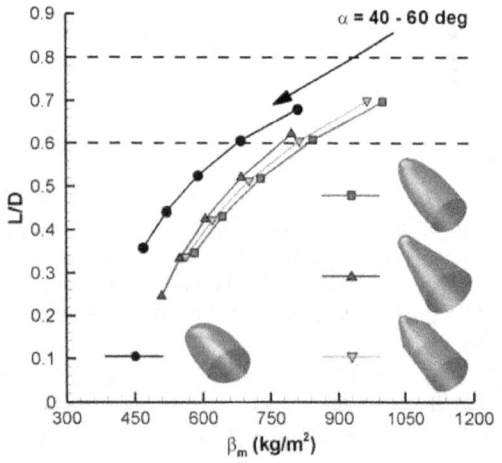

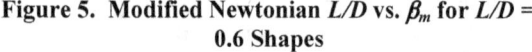

Figure 5. Modified Newtonian L/D vs. β_m for $L/D = 0.6$ Shapes

Shape	L/D	$\beta_m\ (kg/m^2)$	$L\ (m)$	$S_{wet}\ (m^2)$	V_{eff}
	0.608 at 45 *deg*	845	3.14	13.76	0.789
	0.622 at 40 *deg*	796	3.14	14.67	0.717
	0.605 at 45 *deg*	815	3.30	14.09	0.762
	0.606 at 45 *deg*	685	2.36	13.25	0.835

A L/D of 0.8 would give entry corridor margin above 3-σ dispersions. The $L/D = 0.6$ shapes could be flown at angles lower than 40 *deg* in order to improve performance. However, there is a corresponding increase in β_m. Instead, the $L/D = 0.6$ shapes can be lengthened or modified to have a non-circular cross-section. Lengthening of the vehicle would be expected to increase structural requirements and TPS mass. Therefore, modifications to the ellipsled and biconic shapes were studied in order to keep the total orbiter length down while improving L/D to 0.8. Figure 6 and Table 3 summarize those results. The modifications to the ellipsled and biconic both result in $L/D = 0.8$. The longest ellipsled achieves $L/D = 0.8$, but at the expense of large β_m. The ellipsled with an elliptic cross section gave the best combination of L/D and β_m, but the shape is less desirable from a structural standpoint. Bending the biconic nose by 12 *deg* improves L/D to 0.8, but there is a corresponding increase in β_m. The reference orbiter is an ellipsled with a flattened bottom to improve L/D; it is the shortest of all $L/D = 0.8$ shapes and has the highest V_{eff}. Figure 7 and Table 4 summarize the reference orbiter dimensions. The top half of the vehicle has a semi-circular cross-section and the bottom half is elliptical. Other shape classes can be flattened in a similar manner, with a corresponding increase in L/D. Table 5 summarizes estimated mass properties used for the stability analysis. The reference shape was selected for further high-fidelity analyses in the areas of mission design, navigation, aerodynamics, structures, mass properties, aerothermodynamics, TPS, packaging, and guidance.

Table 3. Summary of L/D = 0.8 Shapes

Shape	L/D	β_m (kg/m^2)	L (m)	S_{wet} (m^2)	V_{eff}
	0.806 at 40 deg	1041	3.73	14.48	0.732
	0.799 at 40 deg	694	3.14	14.53	0.728
	0.820 at 40 deg	1116	3.30	14.09	0.762
	0.817 at 40 deg	845	2.88	13.76	0.789

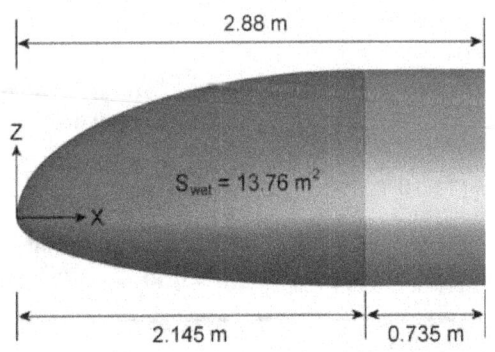

Figure 6. Modified Newtonian L/D vs. β_m for L/D = 0.8 Shapes

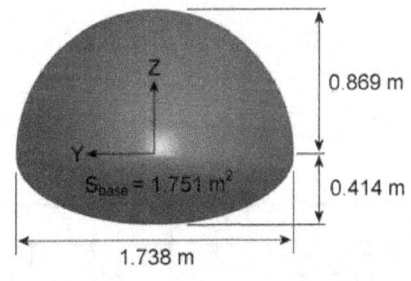

2.88 m

S_{wet} = 13.76 m^2

2.145 m | 0.735 m

0.869 m

0.414 m

S_{base} = 1.751 m^2

1.738 m

Figure 7. Reference Orbiter Geometry

Table 5. Mass Properties

Parameter	Value
m (kg)	2200
I_{xx} $(kg$-$m^2)$	367.8
I_{yy} $(kg$-$m^2)$	857.5
I_{zz} $(kg$-$m^2)$	1035.9
I_{xy} $(kg$-$m^2)$	10.27
I_{xz} $(kg$-$m^2)$	-5.99
I_{yz} $(kg$-$m^2)$	-0.349

Table 4. Reference Parameters

Parameter	Value
L_{ref} (m)	2.88
S_{ref} (m^2)	1.751
X_{cg}/L_{ref}	0.51
Z_{cg}/L_{ref}	-0.0166

STATIC AERODYNAMICS

Viscous CFD solutions were obtained using LAURA at the peak heating point on the minimum atmosphere, lift-up trajectory. Aerodynamic coefficients were calculated for $\alpha = 30 - 50$ *deg* in 5 *deg* increments. The various coefficients are shown in Figure 8 and Table 6. The vehicle trims at 40 *deg* with an axial center of gravity (X_{cg}/L_{ref}) of 0.51 and a lateral CG location (Z_{cg}/L_{ref}) at -0.0166 measured from the waterline. The L/D of 0.806 verifies the original result of 0.817 predicted by Newtonian theory. The ballistic coefficient at trim conditions is 895 kg/m^2. Inviscid non-zero yaw solutions were run using FELISA for use in the lateral stability analysis. Yaw angles of 0, 2, and 5 *deg* were run at $\alpha = 40$ *deg*. Figure 9 and Table 7 summarize those results. The lateral aerodynamics coefficients essentially vary linearly with β for the range of angles examined. The yawing moment coefficient shows stable static behavior ($C_{n_\beta} > 0$). The longitudinal aerodynamics as shown in Figure 8 were included in Monte-Carlo trajectory simulations[4].

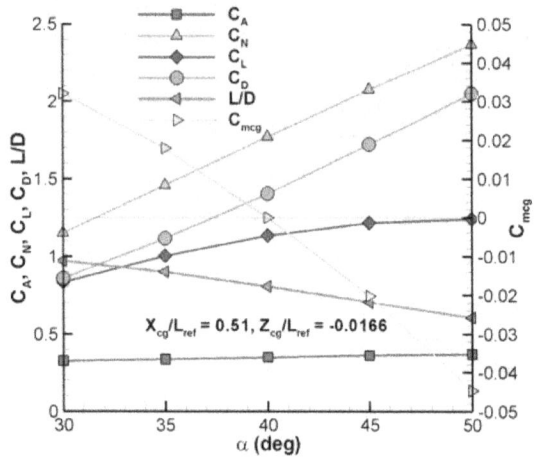

Figure 8. Reference Orbiter Longitudinal Aerodynamics (LAURA Results at $\beta = 0$)

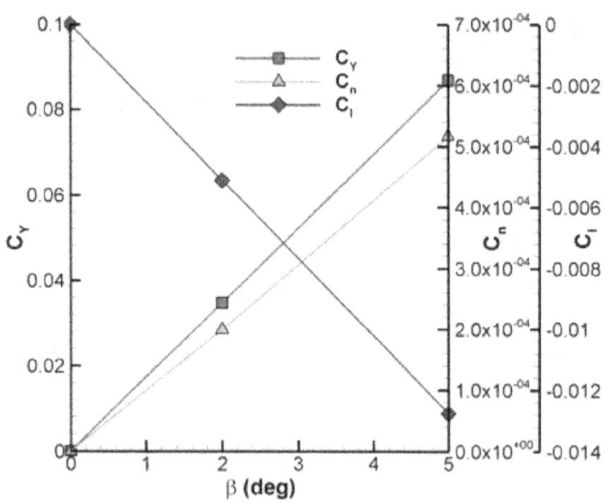

Figure 9. Reference Orbiter Lateral Aerodynamics (FELISA Results at $a = 40$ deg)

Table 6. Reference Orbiter Longitudinal Aerodynamics

α (deg)	C_A	C_N	C_L	C_D	L/D	$C_{m_{cg}}$
30	0.327	1.151	0.833	0.858	0.971	0.0320
35	0.338	1.459	1.002	1.113	0.900	0.0179
40	0.349	1.771	1.133	1.405	0.806	0.0
45	0.359	2.076	1.214	1.722	0.705	-0.0203
50	0.368	2.369	1.241	2.051	0.605	-0.0447

Table 7. Reference Orbiter Lateral Aerodynamics ($\alpha = 40$ deg)

β (deg)	C_Y	C_n	C_l
0	0	0	0
2	3.475×10^{-2}	1.995×10^{-4}	-5.119×10^{-3}
5	8.689×10^{-2}	5.166×10^{-4}	-1.276×10^{-2}

Figure 10 compares the surface pressure coefficient as predicted by the Newtonian method and LAURA CFD on the bottom surface of the orbiter. The excellent agreement between the two methods is reflected in a comparison of the aerodynamic characteristics shown in Figure 11. The Newtonian drag coefficient is within 8% of the LAURA results for the α range between 30 and 50 deg. The difference in predicted L/D is less than 3%. The Newtonian pitching moment coefficient also predicts a trim α of 40 deg and is almost identical to the CFD $C_{m_{cg}}$ results for all α.

The Newtonian method provided remarkably good estimates of hypersonic aerodynamics. This is a reflection of the fact that hypersonic aerodynamics are dominated by the inviscid surface pressure distribution and shear stresses have minimal effect.

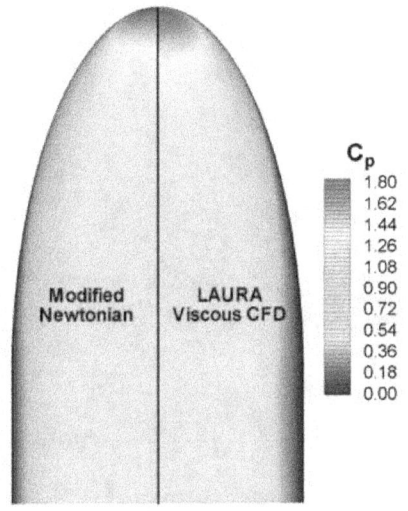

Figure 10. Comparison of Modified Newtonian and LAURA Pressure Coefficient

Figure 11. Comparison of Modified Newtonian and LAURA Aerodynamics

STABILITY ANALYSIS

The reference orbiter C_m shown previously in Figure 8 indicates that the vehicle has a statically stable trim point ($C_{m_a} < 0$) for $\alpha = 40\ deg$. The static margin, shown in Figure 12, is 6.2% of the vehicle's reference length. Typical static margin values for conventional aircraft range from 5% to 45%. This indicates that the reference orbiter has sufficient longitudinal stability and does not compromise the vehicle's longitudinal controllability.

The effect of CG movement on trim can be assessed by specifying the desired trim α and determining the required CG locations to achieve that angle. This analysis results in a line of CG points (the *trimline*) that give zero pitching moment. The trimlines for the reference orbiter at $\alpha = 35, 40,$ and $45\ deg$ are shown in Figure 13. This plot shows the geometrical relationship between CG placement, trim α, and static margin. The angle of each trimline relative to the body axis is determined by $tan^{-1}(C_N / C_A)$ for a given α. Also, each trimline passes through the neutral stability point, at which $C_{m_a} = 0$. Figure 13 shows that there is a small region near $X_{cg}/L_{ref} = 0.51$ where the CG can be realistically placed within the aeroshell. As the CG moves forward or rearward, the required lateral CG eventually moves beyond the outer mold line of the orbiter, and thus is physically impossible.

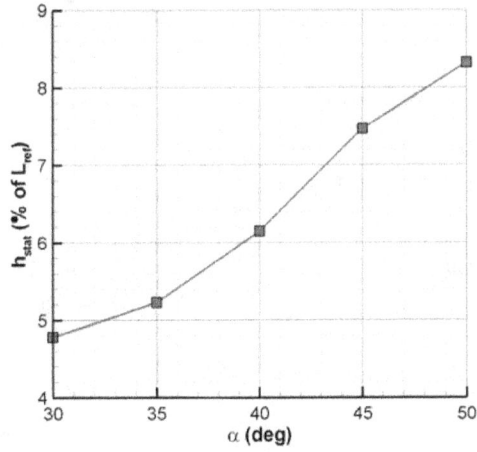

Figure 12. Static Margin vs. α

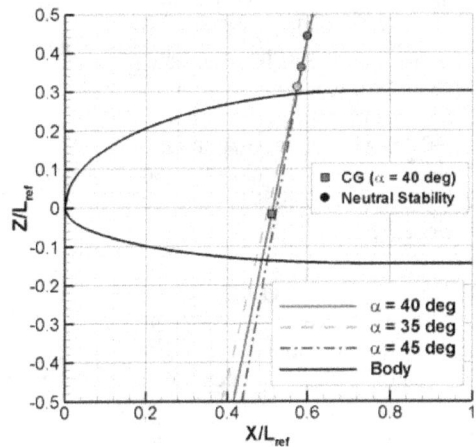

Figure 13. Trimlines as a Function of α

53

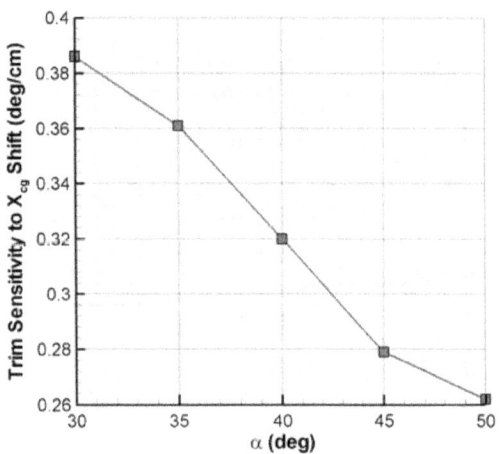 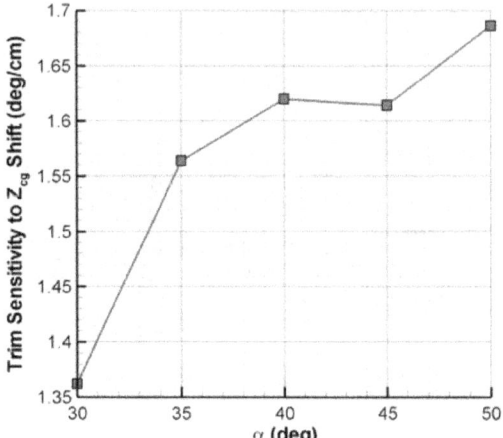

Figure 14. Trim Sensitivity to Changes in Axial (Left) and Lateral (Right) CG Location

Figures 14 shows the sensitivity of trim α to small changes in CG location. These plots show that the trim α is not overly sensitive to CG placement and that trim α is more sensitive to a lateral CG shift than it is to an axial CG shift. This result indicates that the current vehicle design has sufficient margin to accommodate design changes that include a CG shift.

Short–period frequency is shown in Figure 15 and Dutch-roll frequency is shown in Figure 16. The plot in Figure 16 includes the Mil Spec 8785 Level I and II flying quality boundaries (Level I signifies "clearly acceptable" flying qualities, while Level II signifies flying qualities that are "adequate, with some increase in pilot workload")[18]. While these specifications are intended for aircraft flown by human pilots, they do give some indication of how much control effort is required to counter disturbing forces. Given that the response is acceptable for a piloted vehicle, there should be no constraints imposed on an active aerocapture entry control system.

AERODYNAMIC UNCERTAINTIES

Monte-Carlo trajectory simulations were generated for numerous entry conditions and included uncertainty estimates for navigation, atmospheric properties, and aerodynamics[4]. The aerodynamics data based on LAURA solutions and estimated uncertainties were

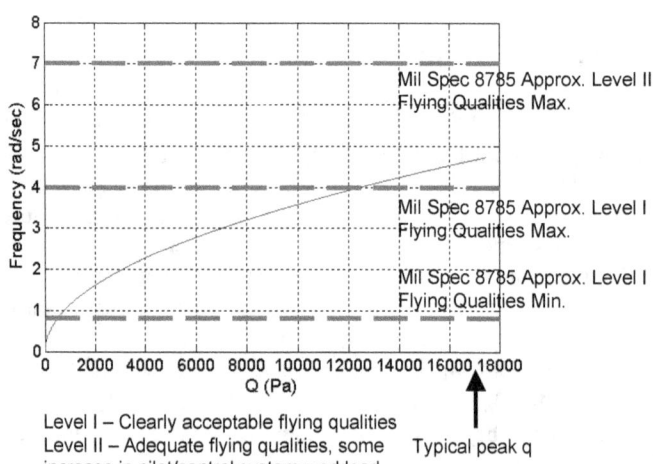

Level I – Clearly acceptable flying qualities
Level II – Adequate flying qualities, some increase in pilot/control system workload

Typical peak q

Figure 15. Short–Period Frequency vs. Dynamic Pressure

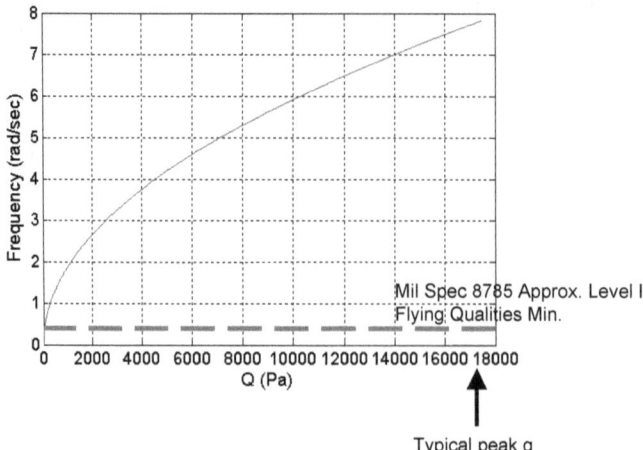

Typical peak q

Figure 16. Dutch–Roll Frequency vs. Dynamic Pressure

supplied for these simulations. The aerodynamics uncertainties are intended to account for computational limitations, shape change due to ablation and deformation, and CG uncertainties. Estimates for aerodynamic uncertainties were based on an ellipsled shape designed for direct entry at Mars[12] and are shown in Table 8 along with CG uncertainties. The aerodynamics uncertainties are consistent with the X-33 uncertainty model[19], which provides a good guideline for mid-L/D shapes. An angle-of-attack uncertainty of 4 *deg* reflects the uncertainty in the pitching moment coefficient and is larger than the uncertainty for typical blunt body entry vehicles. The extreme aerothermal environments[5] could cause significant shape change and alter aerodynamic performance. The effects of large TPS recession on aerodynamic and CG uncertainties have not been quantified.

Figure 17 shows the effects of aerodynamic and CG uncertainties on L/D uncertainty. The largest contributor is the uncertainty in α, followed by the uncertainties in the basic aerodynamic coefficients, C_A and C_N. If the uncertainties are stacked on top of one another, the resulting L/D uncertainty is +26.1%/-22.2%. If the square root of the sum of the squares of the uncertainties is used, the resulting L/D uncertainty is +13.4%/-14.5%. The 3-σ L/D range in a Monte-Carlo analysis would be expected to lie between the stacked and RSS ranges[4].

Table 8. Aerodynamic and CG Uncertainties

Parameter	Uncertainty	Distribution
C_A	+/- 0.048	Uniform
C_N	+/- 0.12	Uniform
α	+/- 4 *deg*	Normal
X_{cg}/L_{ref}	+/- 0.005	Uniform
Z_{cg}/L_{ref}	+/- 0.00125	Uniform

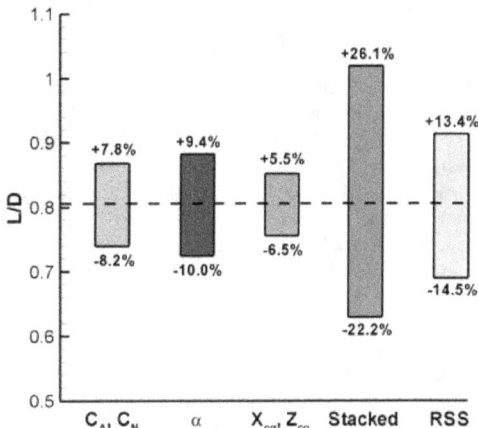

Figure 17. Effect of Aerodynamic and CG Uncertainties on L/D Uncertainty

SUMMARY AND CONCLUSIONS

Aeroshell configuration, aerodynamics, and stability analyses were conducted as part of a systems analysis study for a Neptune orbiter mission. Aerocapture is proposed as the method for delivering the spacecraft to the desired science orbit. Modified Newtonian theory was used to screen candidate mid-L/D shapes for aerodynamic performance ($L/D = 0.6 - 0.8$, minimum β_m) and effective volume (maximum V_{eff}). A flat-bottomed ellipsled with $L/D = 0.8$ and ballistic coefficient of 895 *kg/m²* was selected as the reference orbiter shape. The L/D of 0.8 gives margin to handle cases above 3-σ caused by dispersions in aerodynamics, atmosphere, and navigation. A preliminary assessment was made for static longitudinal and dynamic lateral-directional stability. The orbiter was shown to be longitudinally stable with a 6.2% static margin. It was also determined that the reference orbiter's short-period and Dutch-roll frequencies are acceptable based on piloted-vehicle specifications. Aerodynamics uncertainties were estimated to result in a L/D uncertainty of +13.4%/-14.5% using RSS values and +26.1%/-22.2% using stacked worst-case values.

ACKNOWLEDGMENTS

The authors would like to acknowledge valuable contributions from the following people at NASA Langley Research Center: Brett Starr for generating aerocapture trajectories; Victor Lessard of Geometry Laboratory and Steve Alter for generating the LAURA computational grid; Steve Harris for generating solid models of the candidate orbiter shapes.

REFERENCES

[1] Lockwood, M. K., "Neptune Aerocapture Mission Systems Analysis," AIAA Paper 2004-4951, *AIAA Atmospheric Flight Mechanics Conference*, Providence, RI, 16-19 Aug. 2004.

[2] Justus, C. G., Duvall, A, and Keller, V. W. "Atmospheric Models for Aerocapture Systems Studies," AIAA Paper 2004-4952, *AIAA Atmospheric Flight Mechanics Conference*, Providence, RI, 16-19 Aug. 2004.

[3] Masciarelli, J. P., Westhelle, C. H., and Graves, C. A., "An Aerocapture Guidance Algorithm for the Neptune Orbiter," AIAA Paper 2004-4954, *AIAA Atmospheric Flight Mechanics Conference*, Providence, RI, 16-19 Aug. 2004.

[4] Starr, B. R., Powell, R. W., Westhelle, C. H., and Edquist, K. T., "Performance of a 0.8 L/D Aerocapture Simulation for a Neptune-Triton Exploration Mission," AIAA Paper 2004-4955, *AIAA Atmospheric Flight Mechanics Conference*, Providence, RI, 16-19 Aug. 2004.

[5] Hollis, B. R., Takashima, N., Sutton, K., Wright, M., Olejniczak, J., and Prabhu, D., "Preliminary Convective-Radiative Heating Environments for a Neptune Aerocapture Mission," AIAA Paper 2004-5177, *AIAA Atmospheric Flight Mechanics Conference*, Providence, RI, 16-19 Aug. 2004.

[6] Laub, B., and Chen, Y.-K., "TPS Challenges for Neptune Aerocapture," AIAA Paper 2004-5178, *AIAA Atmospheric Flight Mechanics Conference*, Providence, RI, 16-19 Aug. 2004.

[7] Dyke, R. E., and Hrinda, G. A., "Structural Design for a Neptune Aerocapture Mission," AIAA Paper 2004-5179, *AIAA Atmospheric Flight Mechanics Conference*, Providence, RI, 16-19 Aug. 2004.

[8] Hanak, C. and Bishop, R. H., "Aerocapture Navigation Analysis at Titan and Neptune," AIAA Paper 2004-5180, *AIAA Atmospheric Flight Mechanics Conference*, Providence, RI, 16-19 Aug. 2004.

[9] Tang, W., Orlowski, M., Longo, J., and Giese, P., "Aerodynamic Optimization of Re-Entry Vehicles," *Aerospace Science and Technology*, Vol. 5, Issue 1, pp. 15-25, Jan. 2001.

[10] Wercinski, P. F., Chen, Y.-K., Loomis, M., Tauber, M., McDaniel, M., Wright, M., Papadopolous, P., Allen, G., and Yang, L., "Neptune Aerocapture Entry Vehicle Preliminary Design," AIAA Paper 2002-4812, *AIAA Atmospheric Flight Mechanics Conference*, Monterey, CA, Aug. 2002.

[11] Jits, R. Y., and Walberg, G.D., "High L/D Mars Aerocapture for 2001, 2003, and 2005 Mission Opportunities," AIAA Paper 98-0299, *AIAA Aerospace Sciences Meeting and Exhibit*, Reno, NV, 12-15 Jan. 1998.

[12] Lockwood, M. K., Sutton K., Prabhu, R. K., Powell, R. W., Graves, C. A., Epp, C., Carman, G. L., "Entry Configurations and Performance Comparisons for the Mars Smart Lander," AIAA Paper 2002-4407, *AIAA Atmospheric Flight Mechanics Conference and Exhibit*, Monterey, CA, 5-8 August 2002.

[13] Cheatwood, F. M., and Gnoffo, P. A., "User's Manual for the Langley Aerothermodynamic Upwind Algorithm (LAURA)," NASA TM-4674, April 1996.

[14] Papadopoulos, P., Prahbu, D., Olynick, D., Chen, Y. K., and Cheatwood, F. M., "CFD Code Validation and Comparisons for Mars Entry Simulations," AIAA Paper 98-0272, Jan. 1998.

[15] Hollis, B. R., Thompson, R. A., Murphy, K. J., Nowak, R. J., Riley, C. J., Wood, W. A., and Alter, S. J., "X-33 Aerodynamic Computations and Comparisons with Wind-Tunnel Data," *Journal of Spacecraft and Rockets*, Vol. 38, No. 5, Sept.-Oct., 2001, pp. 684-691.

[16] Peiro J., Peraire J., and Morgan K., "FELISA System Reference Manual and User's Guide," NASA CP-3291, May 1995.

[17] Prabhu, R. K., "Inviscid Flow Computations of Several Aeroshell Configurations for '07 Mars Lander," NASA/CR-2001-210851, April 2001.

[18] Anonymous, "Military Specification: Flying Qualities of Piloted Airplanes", MilSpec 8785C, Aug. 1969.

[19] Cobleigh, B. R., "Development of the X-33 Aerodynamic Uncertainty Model," NASA TP-1998-206544, April 1998.

AEROCAPTURE NAVIGATION AT NEPTUNE

Robert J. Haw

A proposed Neptune Orbiter Aerocapture mission will use solar electric propulsion to send an orbiter to Neptune. Navigation feasibility of direct-entry aerocapture for orbit insertion at Neptune is shown. The navigation strategy baselines optical imaging and VLBI measurements in order to satisfy the flight system's atmosphere entry flight path angle, which is targeted to enter Neptune with an entry flight path angle of –11.6°. Error bars on the entry flight path angle of ±0.55 (3) are proposed. This requirement can be satisfied with a data cutoff 3.2 days prior to arrival. There is some margin in the arrival template to tighten (*i.e.* reduce) the entry corridor either by scheduling a data cutoff closer to Neptune or alternatively, reducing uncertainties by increasing the fidelity of the optical navigation camera.

INTRODUCTION

An orbiter mission is described combining solar electric propulsion for an inter-planetary transfer to Neptune and aerocapture technology for orbit insertion upon reaching Neptune. This paper evaluates the feasibility of navigating a direct-entry aerocapture at Neptune. The work is part of a NASA inter-center study [Ref 1].

Aerocapture is an orbit insertion flight maneuver within a planetary atmosphere using drag to decelerate the spacecraft to orbital velocities with a single pass through the atmosphere. It requires zero or minimal propellant to effect the orbit insertion. Aerocapturing into a closed elliptical orbit around Neptune has the advantage of allowing higher entry velocities than would otherwise be possible, thus reducing the interplanetary transfer time. It also reduces arrival mass for a given payload mass.

An established accuracy requirement for the navigation sub-system did not exist at the time of this study. One of the purposes of this work then, was to set limits on the navigation error and, in collaboration with aerocapture colleagues [Ref 1], determine aerocapture accuracy requirements for navigation.

An error analysis requires detailed inputs in order to build the navigation model, so first a representative spacecraft proposed by the aerocapture study group is described [Ref 2 & 3], followed by a description of the target selection and mission design. Several system trades are subsequently performed, including a trade on entry velocity at Neptune.

REPRESENTATIVE SPACECRAFT CONFIGURATION

The orbiter is enclosed by an aeroshell. The mass of the entire entry flight system is 1800 kg (including propellant load). The lift-to-drag ratio (L/D) of the vehicle equals 0.6 and its ballistic coefficient (M/C$_D$A) is approximately 150 kg/m^2.
The spacecraft is modeled as a 3-axis-fixed spacecraft with momentum-wheel ACS stabilization. The momentum wheels maintain spacecraft pointing, and balanced thrusters perform periodic momentum de-saturation burns.

Solar electric propulsion (SEP) boosts the package after launch. SEP thrusts within the inner solar system, but all engines and solar arrays are discarded beyond ~3 a.u. A mono-propellant hydrazine propulsion system remains after jettisoning the SEP. This subsystem performs the momentum wheel de-saturations and trajectory correction maneuvers (TCMs) during the approach to Neptune.

The telecommunications sub-system during the interplanetary transfer employs a Ka (or possibly X) -band high gain antenna (HGA) mounted on the back of the aeroshell for telemetry and navigation. The HGA is a 1 m diameter dish antenna, with a 5 watt transmitter and a gain of 36dBi. At 30 a.u. the data rate to a 70 m ground antenna is ~100 bps. Also mounted on the aeroshell are forward-looking cameras for optical navigation.

The tracking and telemetry sub-system will use a Small Deep Space Transponder, which supports phase coherent two-way doppler and ranging, command signal demodulation and detection, telemetry coding and modulation, and differential one-way range (DOR) tone generation (for ΔVLBI measurements).

TARGET SELECTION

The Neptune target is determined by the post-insertion orbit, atmosphere characteristics, and the aerodynamic performance of the entry vehicle.

The entry interface (EI) target at the top of the atmosphere consists of three parameters: inertial flight path angle (FPA), clock angle, and radius. The flight path angle is the angle subtended by the vehicle trajectory with the local horizontal at the entry interface radius (see Appendix 1). The FPA used for this study is -11.6° [Ref 1]. The clock angle, as its name suggests, is a clockwise angular measure of the position of the target point on the projected face of Neptune's disk, measured from the **T** axis (see Appendix 2). The entry interface radius is defined to be 25,757.0 km (an altitude 1000 km above the 1 bar ambient pressure level) [Ref 1].

The entry interface target and desired orbit characteristics are provided in Table 1.

Table 1. Entry Interface Target And Post-Insertion Orbit Characteristics
Entry time: 2021 April 28 00:09 UTC

EI Target (Retrograde Entry)				Initial Orbit Characteristics		
Altitude (km)	Latitude (deg)	eFPA (deg)	Entry Velocity (km/s)	Altitude (km)	Inclination (deg)	Period (hours)
1000	7N	-11.6	28.0	4000* x 430,000	157	80

* After the pericenter-raise maneuver

TRAJECTORY OVERVIEW

A pair of Neptune trajectories was chosen to perform trades. The pair is representative of a single interplanetary trajectory found by Sauer and Noca [Ref 4]. The two trajectories are equivalent except for their hyperbolic excess velocity. The v_∞'s are 15.9 km/s and 18.5 km/s.

The vehicle arrives at Neptune on April 28, 2021 after a journey of 10 or more years. The range to Earth at entry is 29.8 a.u. (one-way light time equals 4 hours).

The approach trajectory is ballistic. The entry is retrograde, making the atmosphere-relative velocity at the EI significantly higher than for a prograde entry. A retrograde entry imposes rigorous requirements on the flight system (high decelerations and heat loadings) but the subsequent capture orbit is preferable because it facilitates rendezvous with Triton (inclination equal to 157° and orbit period of 5.9 days). Alternatively, as shown in Reference 5, a prograde entry is less demanding on the flight system but penalizes the mission with long period orbits (on the order of months) and expensive inclination changes (to align the spacecraft with Triton's orbit). The retrograde option was chosen for this study.

Reference 5 recommends an entry velocity near 29 km/s. Trades are performed here on entry velocities of 28 km/s and 30 km/s, corresponding to v_∞= 15.9 km/s and 18.5 km/s respectively.

The entry vehicle enters the atmosphere at an altitude of 1000 km above the 1 bar level and descends to ~200 km before climbing and exiting the atmosphere. Near apocenter, a pericenter-raise maneuver is performed to raise pericenter out of the atmosphere. See Figure 1. The Δv needed to raise pericenter to an altitude of 4000 km is 90 m/s. A 4000 km pericenter lies well above the atmosphere and satisfies Neptune–science measurement objectives [Ref 6].

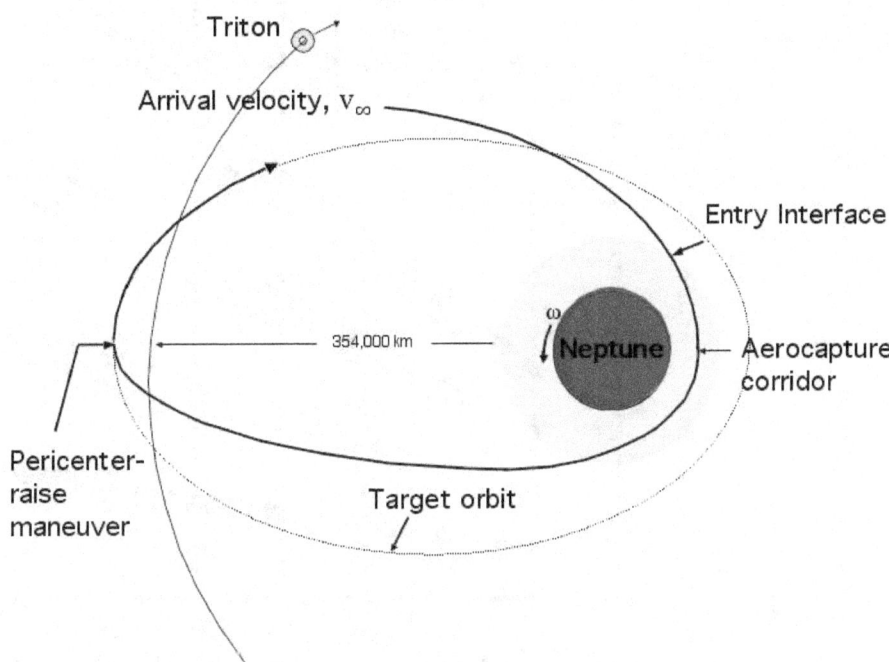

Figure 1   Schematic of Neptune Aerocapture

An entry velocity of 28 km/s decelerates the orbiter with a unit force of ~5±3 g's and slows the vehicle ~6 km/s by the time of egress. At 30 km/s the mean deceleration is ~8 g's [Ref 3].

Short period orbits around Neptune (days, rather than months) are desirable for science objectives. In addition to yielding more science quantitatively, short period orbits have relatively low hyperbolic velocities -- an advantage for tour design and for observing Triton (assuming apocenter is greater than Triton's orbit). Tour design benefits because gravity assist swing-bys are more efficient at low velocities (inverse-square relationship between trajectory-bending and v_∞). Triton observations benefit because longer exposure times are possible. Therefore Triton approach velocities should be kept as low as practical.

The orbit radius of Triton is 354,000 km. Using the previous considerations as guidance, an apocenter equal to 430,000 km was selected to satisfy science requirements and orbit commensurability. This distance defines an orbit period of 80 hours and a 7:4 resonance with Triton [Ref 7].

AEROCAPTURE

Entry flight path angle is constrained by the physical limitations of the flight system (the vehicle must withstand aerodynamic, structural, and heat loads), and by the need to accumulate sufficient drag forces to slow the spacecraft (to avoid skipping-out). Error bars on the entry trajectory define a corridor through the atmosphere.

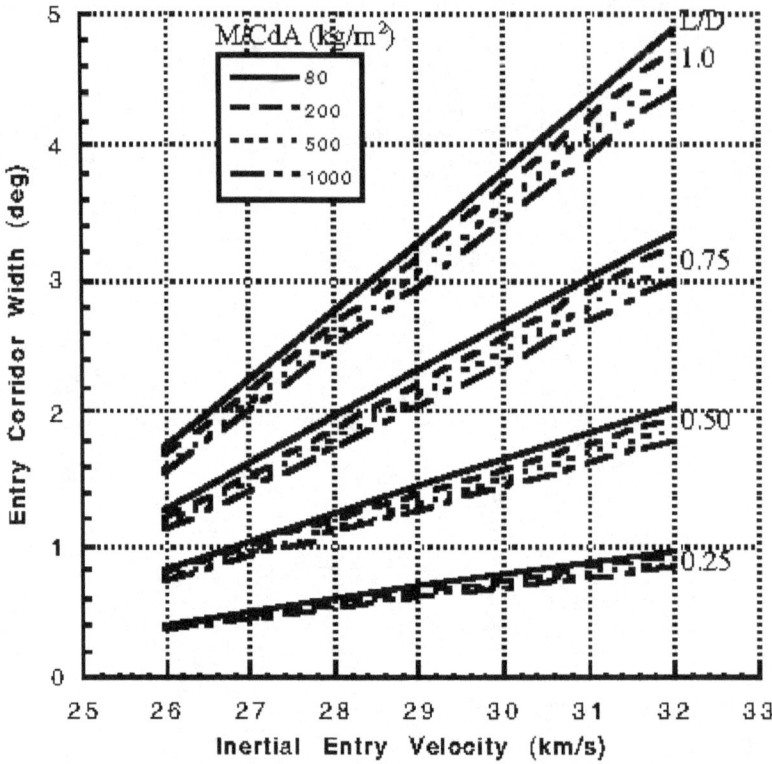

Figure 2   Variation of Entry Corridor Width as a Function of Entry Velocity and Ballistic Coefficient. Apocenter = 400,000 km [Ref 5]

The diameter of the corridor through the atmosphere represents the maximum total uncertainty that can be accumulated by the vehicle -- with contributions from the atmosphere, vehicle aerodynamics, and navigation. Corridor diameter defines the maximum tolerable limits along the aerocapture flight path.

Reaching the desired apocenter then, is a function of adjusting entry velocity, L/D, and ballistic coefficient. In Figure 2 entry corridor width is plotted versus entry velocity for a range of ballistic coefficients bounding representative vehicle sizes. (Note that this figure applies to an apocenter radius of 400,000 km and not 430,000 km.) For the vehicle used here and an entry velocity of 28 km/s, Figure 2 indicates a theoretical corridor width of ~1.4° (*i.e.* ±0.70°) while an entry velocity of 30 km/s specifies a theoretical width of ~2.0° (*i.e.* ±1.0°). Higher entry velocities provide additional margin but subject the vehicle to greater stress.

The errors contributing to a corridor width corresponding to an entry velocity of 28 km/s and a vehicle L/D of 0.6 are shown in Figure 3. The abscissa represents atmosphere variability, where the dimensionless parameter Fminmax varies from minimum atmospheric density (-1) through maximum density (1).

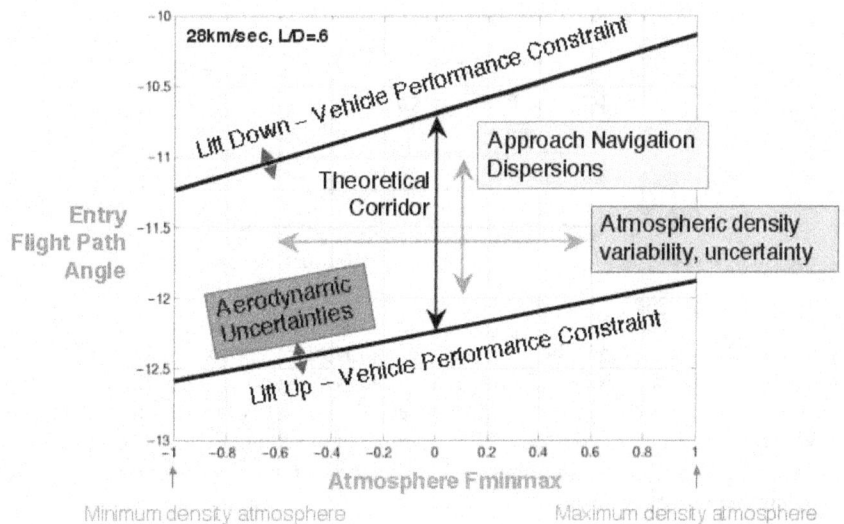

**Figure 3 Vehicle Performance for Apocenter = 430,000 km
E-3 day Delivery [Ref 3]**

Navigation dispersions contribute approximately 77% of the corridor uncertainty in Figure 3 for Fminmax equal to –1 (the narrowest point). Conservatively assuming the same proportional contribution in the center of the plot at Fminmax equal to 0, a flight path angle requirement appropriate to levy on navigation will be 77% of 1.4° or a 3σ error of approximately ±0.55° for an entry velocity of 28 km/s, and a 3σ error of approximately ±0.75° for an entry velocity of 30 km/s. (Note that the corridor width in Figure 3 defines 100% values although it has been asserted here that the limits represent 3σ values. This equivalence is inconsistent but is not troubling because the margins for error in this analysis are greater.)

NAVIGATION DATA

The navigation accuracy achievable at the destination, or target, is established at the final control point along a trajectory (*i.e.* the last maneuver before reaching say, Neptune) and is usually termed the delivery accuracy, or simply the *delivery*. Since there is a limit on the accuracy with which an initial state and subsequent dynamics are known, the future state cannot be computed with complete certainty from an initial one. So a delivery at time T includes the future uncertainty expected in the spacecraft state (at its time of arrival♦) computed at time T, where T is before the time of arrival. That is, an E–2 day delivery represents the prediction of the location of the spacecraft at Neptune, when still 2 days away from Neptune.

The error analysis undertaken here begins at Neptune-90 days.

EPHEMERIS DETERMINATION

Ephemeris errors dominate the navigation errors at Neptune and an aerocapture mission isn't feasible unless these errors are reduced significantly. Neptune's ephemeris errors, as well as Earth's, are

---

♦ More specifically, 'entry time', defined in Table 1.

given in Table 2. This tabulation is given in a Sun-centered RTN coordinate system, where R represents radial direction from the Sun, T down-track direction (the direction of motion of Neptune in its orbit), and N the out-of-plane direction.

Table 2. Neptune And Earth Ephemeris Uncertainties (3σ)
Mapped to 2021

Central Body	R (km)	DT (km)	OOP (km)	TOTAL (RSS)
Earth	0.01	3	4	5
Neptune*	10,200	12,000	5,200	16,000
Neptune**	3,400	4,000	1,733	5,200

* DE405 (circa 2000)
** Uncertainties used for this analysis

The second line in Table 2 represents the error in Neptune's position in 2021 as currently projected by JPL's DE405 planetary ephemeris (*i.e.* a mapping of 21 years). (In an absolute sense these errors are large, but the total error is only about one-third the planet's diameter.) Significant improvements to the ephemeris between now and 2021 can be expected. As more observations are acquired between now and arrival, *a priori* errors will decrease. For example, the Neptune *a priori* ephemeris error for Voyager II was ~5000 km RSS (3σ). For this analysis the assumed *a priori* error are one-third the DE405 errors. Those errors are shown on the third line in Table 2.

OPTICAL DATA

Target-relative imaging is important in this mission because of the uncertainty in the location of Neptune. Optical navigation data are used to reduce Neptune's errors. These data consist of digital images of Neptune and its satellites, set in front of a stellar background. The background stars, combined with Neptune's ephemeris, establish the spacecraft-Neptune relative position by astrometry.

The optical navigation campaign begins at E-75 days. Ground-based facilities will process the transmitted pictures to extract the optical observables, and the data will be combined with radiometric measurements. Data processing and observable-extraction require approximately eight hours to complete (as of 2003).

Transmissions will be constrained by the down-link data rate (~100 bps). A schedule of one image per every four hours (6 pictures per day) satisfies this constraint.

Early in the approach phase, one picture every other day is shuttered, alternating between Neptune and Triton. The picture frequency increases to six per day within 16 days of Neptune. This yields approximately 170 images in the complete optical data set.

The imaging system envisioned here follows a design similar to the Mars Reconnaissance Orbiter optical navigation camera. Relevant technical specifications of the MRO camera are: aperture = 6 cm, focal length = 50 cm, field-of-view = 1.4° per side, detector = 1024x1024 CCD array, pixel resolution = 50 μrad, mass = 2.7 kg, peak power = 4 W [Ref 8]. Higher performance cameras will yield better results. Therefore an advanced camera ("MRO plus") with a pixel resolution of 40 μrad is also parameterized to show relative performance.

For comparison with an operating mission, the Cassini wide-angle navigation camera has these specifications: aperture = 6 cm, focal length = 20 cm, field-of-view = 3.5° per side, detector = 1024x1024 CCD array, pixel resolution = 60 μrad, mass = 27 kg, peak power = 35 W [Ref 9]. The MRO camera offers higher resolution than Cassini, yet weighs less and operates with less energy (but has a smaller field-of-view).

See Appendix 3 for other camera parameters.

TRACKING DATA

Navigation tracking data consists of two-way and three-way coherent Ka-band doppler and range measurements. (X-band data were found to perform equally as well.) These data are augmented during approach with optical observations and interferometry.

Interferometry enhances the solution relative to that achievable with doppler, range and optical data (although optical data dominates in a ranking of the relative importance of the four data types). In general though, interferometric data *i.e.* Delta Differenced One-way Range (ΔDOR), has limited effectiveness at the range of Neptune, although it can be used to some advantage in combination with the other data types. That is, while optical data determines plane-of-sky information for Neptune (from which the plane-of-sky position of the spacecraft can be inferred), ΔDOR measurements can determine plane-of-sky spacecraft components directly.

ΔDOR measurements are not constrained by the down-link, but require 24 hours to extract the observable from the data (conservatively).

Data schedules used in this analysis for doppler, range and ΔDOR are provided in Table 3. Data measurement accuracies are listed in Appendix 3.

Table 3. Doppler & Range Tracking And ΔDor Measurements

Start	End	Radiometric Coverage	Start	End	ΔDOR Observations
E-90days	E-60	2 tracks/week			
E-59	E-45	1 track /day	E-75days	E-51	1 per week
E-44	E-17	2 tracks/day	E-50	E-31	3.5 per week
E-16	Entry	3 tracks/day	E-30	Entry	14 per week

NAVIGATION MODEL

Significant error sources in the navigation model are noted in the sub-sections below. Appendix 3 lists all error sources and *a priori* uncertainties.

MANEUVER PLACEMENT

Maneuvers during the approach phase were placed as shown in Table 4 below. This is a representative schedule put together for the purposes of the error analysis. The last targeting maneuver during approach is the most sensitive to placement, for it defines the delivery accuracy. For this reason two opportunities are shown in Table 4 for the final targeting maneuver: E-2 days and E-1 day. For the baseline strategy (E-2 days) the data cutoff is 3 days from Neptune, whereas the alternative strategy proposes a data cutoff 2 days from Neptune. The alternative strategy delivers smaller uncertainties but leaves less time to correct those errors before entry.

Table 4. NEPTUNE ORBITER Tcms

TCM*	Time**	Data Cutoff**	Description
TCM1	E -60 days	E - 65 days	Correct SEP cruise errors.
TCM2	E -10 days	E – 15 days	Penultimate targeting
TCM3	E – 2 days	E – 3 days	Ultimate targeting
TCM3'	E – 1 day	E – 2 days	Ultimate targeting (alternate)
TCM4	~E +0.75 day	~E + 0.1 day	Apocenter correction
TCM5	~E +1.75 day	~E + 0.85 day	Pericenter-raise to 4000 km

*Numbered starting at the beginning of the approach phase.
**With respect to entry (E) time.

All maneuvers in Table 4 except TCM5 are statistical maneuvers. The statistical analysis necessary to size the statistical maneuvers has not been performed, but the mean Δv for TCM1 and TCM2 probably will not exceed 1 m/s (based on the ephemeris errors). The expected Δv for either TCM3 or TCM3' will be greater (but it is not expected to be more than an order of magnitude greater). The deterministic component of the TCM5 magnitude is 90 m/s.

ORBIT DETERMINATION

The dominant orbit determination uncertainties consist of ephemeris errors, TCM execution uncertainties, and data errors. The uncertainties contributing to orbit determination errors are listed in Appendix 3.

RESULTS

Delivery errors are a combination of orbit determination errors and maneuver execution errors, mapped to the entry interface. Sensitivity trades in this sub-section look at optical navigation and/or ΔDOR observations, delivery time, ephemeris errors, and entry velocity.

Delivery uncertainties are plotted in Figure 4 below. Neptune's ephemeris uncertainty is the predominant reason for the large uncertainties at the left edge of the figure.

Figure 4 illustrates six options, or strategies. Option 1 is the baseline case: *i.e.* MRO-like camera with 6 pictures per day maximum, plus doppler, range and ΔDOR.

Option 2 doubles the number of pictures acquired by the camera during the last two weeks of approach (an unlikely scenario given the assumed down-link rate). This option shows appreciable improvement with respect to Option 1. Option 3 suggests the benefits that an advanced camera (MRO-plus) may offer. Its performance (with 6 pictures per day) is equivalent to Option 2 (with 12 pictures per day). Option 4 lacks ΔDOR measurements. Some degradation occurs with this loss but the degradation is not significant. The loss of ΔDOR can be balanced by substituting the advanced camera. Option 5 illustrates the performance of the Cassini wide-angle camera. It does not perform as well as the baseline case. Option 6 illustrates the inappropriateness of performing this mission with only doppler and range data.

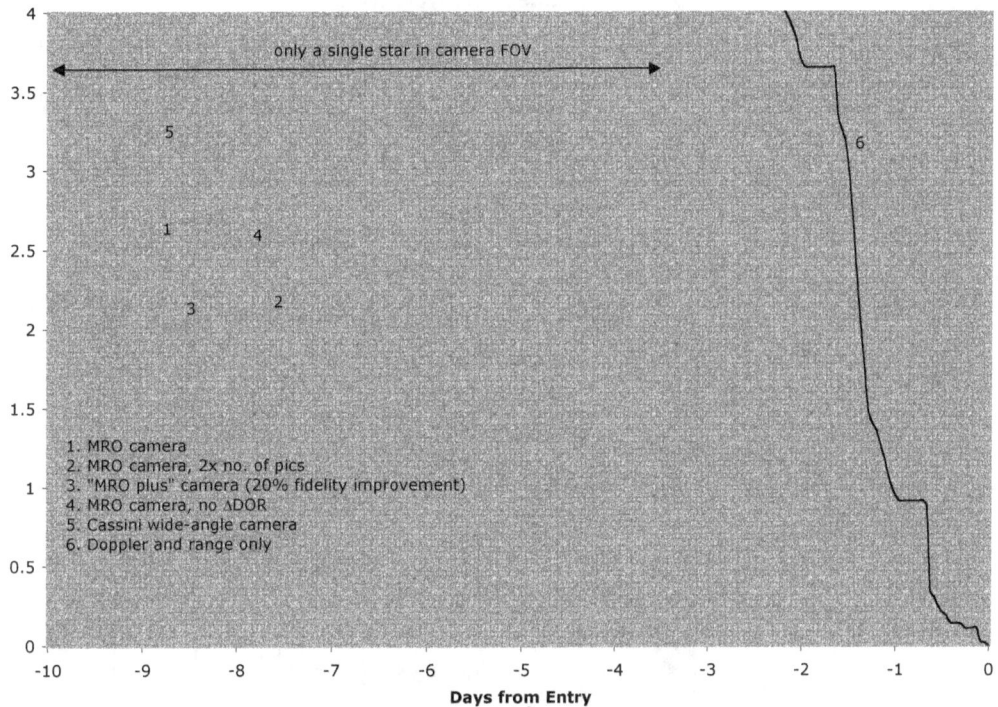

Figure 4 FPA v. Time-to-Go (3σ)
Entry Velocity = 28 km/s

The geometry of this trajectory lacks bright background stars suitable for optical navigation (*i.e.* few are visible behind Neptune). The navigation information content per image is enhanced with multiple background stars, but a narrow field of view reduces the probability of capturing more than a single bright star. (Multiple stars determine the center point of Neptune more accurately because of the additional degrees-of-constraint introduced.) The camera performances shown in Figure 4 are not optimum because only a single star is visible per image until E-3.5 days. At that time a second star enters the field-of-view, and the improvement in the delivery is significant.

Sensitivities are shown in Tables 5 - 8. The entries in Tables 5 and 6 reproduce Figure 4 in greater detail at selected times. Note that FPA entries in these tables don't mirror values from Figure 4 at those selected times. The processing time for computing optical and ΔDOR observables introduces a lag during flight operations. That lag has been accounted for in Tables 5 and 6, but is not computed in Figure 4. That is, Figure 4 represents instantaneous processing of optical and ΔDOR measurements. For Tables 5 – 8, processing delays of 10 hours for optical data and 24 hours for ΔDOR are assumed.

Tables 5 and 6 show that the proposed delivery requirement is satisfied at the time of Delivery B (E-3 days) for all of the tracking options with cameras. (Except for the Cassini option which narrowly misses.)

Table 5. Delivery Accuracy (3σ)
28 KM/S

	Doppler & Range Only	Doppler Range & Optical	Doppler Range Optical ΔDOR	Doppler Range Optical ΔDOR
Proposed Reqm't	±0.55	±0.55	±0.55	±0.55
Delivery A				**2x pics**
Data Cutoff at E-4.25 days		145 pics	145 pics	213 pics
Semi-major axis (km)	1152	234	222	183
Semi-minor axis (km)	591	84	57	51
Ellipse angle (deg)	67	21	24	22
Entry time (s)	117	36	33	27
B magnitude (km)	720	222	217	171
Flight Path Angle (deg)	**±4.5**	**±1.5**	**±1.3**	**±1.1**
Delivery B				**2x pics**
Data Cutoff at E-3 days		153 pics	153 pics	229 pics
Semi-major axis (km)	1122	84	81	60
Semi-minor axis (km)	588	63	48	42
Ellipse angle (deg)	68	35	23	20
Entry time (s)	114	12	12	9
B magnitude (km)	702	78	78	60
Flight Path Angle (deg)	**±4.4**	**±0.48**	**±0.48**	**±0.37**
Delivery C				**2x pics**
Data Cutoff at E-2 days		159 pics	159 pics	241 pics
Semi-major axis (km)	1011	57	45	36
Semi-minor axis (km)	534	39	36	27
Ellipse angle (deg)	76	65	49	59
Entry time (s)	102	6	6	6
B magnitude (km)	582	42	39	30
Flight Path Angle (deg)	**±3.7**	**±0.27**	**±0.24**	**±0.18**
Parameter Update				**2x pics**
Data Cutoff at E-12 hours		168 pics	168 pics	258 pics
Semi-major axis (km)	885	36	33	27
Semi-minor axis (km)	30	18	18	15
Ellipse angle (deg)	93	102	98	98
Entry time (s)	3	3	3	2
B magnitude (km)	33	18	18	15
Flight Path Angle (deg)	**±0.21**	**±0.11**	**±0.11**	**±0.09**

Table 6. Camera Sensitivity (3σ)
28 KM/S

	MRO camera (baseline)	MRO camera (baseline) 2x pics	MRO plus camera	Cassini camera
Proposed Reqm't	±0.55	±0.55	±0.55	±0.55
Delivery B				
Data Cutoff at E-3 days	153 pics	229 pics	153 pics	153 pics
Semi-major axis (km)	81	60	66	153
Semi-minor axis (km)	48	42	42	75
Ellipse angle (deg)	23	20	23	23
Entry time (s)	12	9	9	24
B magnitude (km)	78	60	63	144
Flight Path Angle (deg)	**±0.48**	**±0.37**	**±0.39**	**±0.89**

Table 7 shows the effect of improvements to the *a priori* Neptune ephemeris. There are no significant improvements. That is, the delivery is not sensitive to ground-observation updates to the ephemeris *i.e.* the current planetary ephemeris DE405 is satisfactory. This is an unexpected result, but verifies the value of the optical navigation.

Table 7. Entry Fpa -- Ephemeris Sensitivity (3σ)
28 KM/S

Neptune ephemeris ->	Baseline circa 2021	DE405 (mapped to 2021)
Data Cutoff		
E – 3 days (deg)	±0.48	±0.49
E – 2 days (deg)	±0.24	±0.25
E – 12 hours (deg)	±0.11	±0.11

Table 8 shows that entry flight path angle uncertainty is proportional to entry velocity, as expected.

Table 8. Entry Fpa -- Entry Velocity Sensitivity (3σ)

Data Cutoff	28 km/s	30 km/s
E – 3 days (deg)	±0.48	±0.79
E – 2 days (deg)	±0.24	±0.36
E – 12 hours (deg)	±0.11	±0.20

Flight path angle dispersions shown in Table 8 are plotted versus entry velocity in Figure 5.

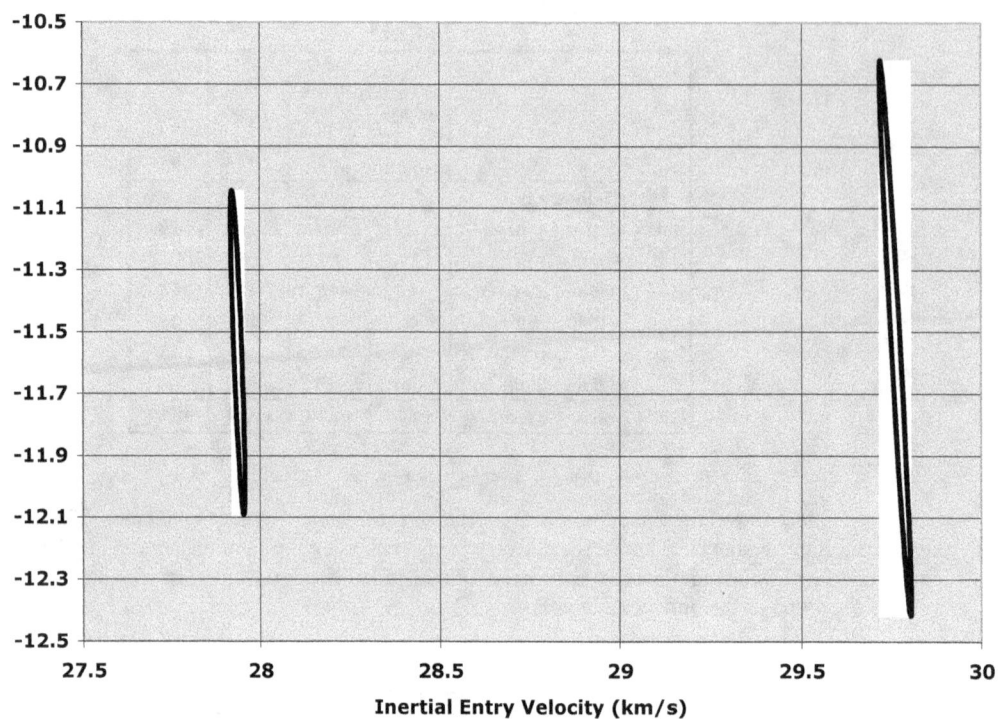

Figure 5   Navigation Dispersions for 28 km/s and 30 km/s at EI (99%)
E-3 day Delivery

DISCUSSION

OPTICAL NAVIGATION DATA

This mission cannot be performed without optical navigation, and even optical navigation offers little margin. For the nominal case under consideration (MRO-like camera, picture frequency = 6 per day), the delivery requirement is satisfied at about E-3.2 days. An additional day can be purchased, *i.e.* delivery at about E-4 days, by employing a camera more advanced than the MRO version (MRO-plus) or by increasing the downlink rate to support a higher picture frequency. This strategy yields modest improvements, and the sensitivity to further camera development is evident.

A more likely source to find immediate additional paper margin is from a well-designed picture sequence command file. The picture sequence file used in this analysis captured multiple stars only for the last 3.5 days, and the effect (the difference between a single and multiple stars) is significant and self-evident (see Figure 4 again). With detailed optical navigation planning the delivery requirement could be satisfied by ~E-4.5 days for the nominal case, or as early as ~E-5 days for the advanced camera (by extrapolation in Figure 4).

Navigation images of Neptune, because of its atmosphere, have relatively large uncertainty (especially during the two weeks preceding entry). This uncertainty was mitigated by incorporating pictures of Triton because airless bodies do not degrade optical data in a way that an atmosphere does. (A ratio of 2 Triton pictures for every 1 Neptune picture was used.)

The *a priori* ephemeris of Neptune is not important to the delivery. The mission can be undertaken with the current DE405 ephemeris and the current Triton *a priori* ephemeris.

TRACKING DATA

ΔDORs and optical data are orthogonally complementary and combine to yield plots 1, 2, 3 and 5 in Figure 4. Plot 3 assumes an advanced camera (MRO-plus) and represents the best delivery in the current study (but <7% improvement over the baseline). Note that plot 2 is similar to plot 3, but represents a less advanced camera shuttering at twice the frequency.

ΔDOR measurements improve delivery accuracy <5% after accounting for the data processing lag (instantaneously the improvement is ~10%). Improvement is possible because ΔDOR observations are sensitive to state errors along components insensitive to doppler and range.

There is no advantage to using Ka-band doppler tracking in place of X-band. Small benefits were seen with Ka-band ΔDOR observations (*vis-a-vis* X-band observations), but the overall improvement to the delivery was insignificant.

PROPELLANT BUDGET

As a rough estimate of propellant loading, at least 105 m/s of velocity change is required to get into orbit (*i.e.* not including on-orbit maintenance propellant nor the allocation necessary to perform a Neptune-Triton orbital tour). The 105 m/s total is composed of ~12 m/s of pre-insertion statistical Δv and a deterministic Δv =90 m/s for TCM5 (the pericenter-raise maneuver – note: the statistical component of this maneuver is still TBD). An additional statistical maneuver (TCM4) is needed between egress and apocenter (before TCM5) to correct residual aerocapture errors and achieve the apocenter target. The size of this maneuver is TBD. The 105 m/s total is expected to grow significantly with these TBD additions.

TCM4 and TCM5 are scheduled with only one day separating them, and both maneuvers must be designed and burned within ~40 hours of egress. This is a difficult, but not impossible task to accomplish using traditional maneuver template procedures (*i.e.* no autonomy).

COMPARISON WITH OTHER MISSIONS

Entry FPA results (or expected results) from other missions are summarized in the table below. (MER, Stardust, and Huygens have not yet arrived at Mars, Earth, and Titan respectively at the time of this writing.)

Table 9. Fpa Delivery Comparison (3σ)

Mission	Entry FPA	Delivery Error	Delivery Time	Reqm't
Neptune Orbiter	-11.6°	±0.24°	E-2 d	<±0.55>
Titan Explorer*	-36.8°	±0.6°	E-2 d	<±1.0>
Mars Pathfinder	-14.2°	±0.4°	E-2 d	±1.0
MPL	-12.0°	±1.0°	E-2 d	~±0.5
MER	-11.5°	±0.2°	E-2 d	±0.25
Stardust	-8.2°	~±0.8°	E-2 d	±0.80
Galileo probe	-8.6°	±0.6°	E-140 d	±1.4
Huygens probe	-64.0°	±3.0°	E-21 d	±3.4

* Proposed mission. See Reference 10.
<-> denotes proposed requirement.

MPL and Stardust stand out in the short list above with high uncertainties.

The MPL mission was characterized by unbalanced and mis-modeled thrusting activities. The level of thrusting required by the ACS system to maintain attitude significantly exceeded pre-launch expectations, and this mis-modeling contributed to the entry flight path angle uncertainty shown in Table 9.

Huygens (the Cassini probe) anticipates a delivery uncertainty of ±3.0°. One reason for the large delivery error is the Titan ephemeris uncertainty. Another reason is the tour re-design Cassini has undergone recently [Ref 11].

The MER delivery, on the other hand, is significantly smaller than the Neptune Orbiter delivery. Mars' well-known ephemeris and MER's lower hyperbolic excess velocity are contributors to this improvement.

SUMMARY

This preliminary study has baselined the use of optical observations and ΔDOR measurements for delivering an aerocapture orbiter to Neptune. The study has also proposed a conservative entry FPA requirement of ±0.55° (3σ) based on delivery results that accommodate the aerocapture. The proposed delivery requirement is satisfied at E-3.2 days. This date can be pushed earlier in all likelihood (further from Neptune) with subsequent follow-up optimization of (i) the picture sequence file and (ii) camera design.

This study makes two recommendations to enhance performance at Neptune:

- Development of a targetable, advanced optical navigation camera. The MRO navigation camera currently under development represents a satisfactory technological readiness level, but an advanced version will buy margin.
- Second, incorporation of on-board autonomous maneuver capability.

ΔDOR measurements offer negligible benefit. This analysis does not support a navigation strategy incorporating ΔDOR measurements.

Proposed entry requirements can be met using the equivalent of a future DE405 *a priori* Neptune ephemeris such as that described in Table 2.

This work represents a first-cut effort at determining concept feasibility. Many simplifying assumptions were made, especially with respect to the optical data, in order to accomplish this study in a timely manner.

ACKNOWLEDGEMENT

The work described in this paper was performed at the Jet Propulsion Laboratory, California Institute of Technology under contract from the National Aeronautics and Space Administration. The work was funded through the In-space Propulsion program.

The author would like to thank D.W. Way, M.K. Lockwood, B. Starr, J.L. Hall, W.M. Owen, and L.A. Cangahuala for helpful comments and support in preparing this paper. Carl Sauer constructed the intial SEP trajectory between Earth and Neptune.

REFERENCES

1. M.K. Lockwood, Langley Research Center. Team lead, "Neptune aerocapture systems analysis study group", NASA In-space Propulsion program, Oct. 2002 – Aug. 2003.
2. R.W. Bailey, JPL, private communication, Dec 2002.

3. M.K. Lockwood, D.W. Way, B. Starr, Langley Research Center, private communication and viewgraph presentations, Dec 2002 – April 2003.
4. C. Sauer, M. Noca, JPL, private communication, Sept 2002.
5. P.F. Wercinski, Y-K Chen, M. Loomis et al, "Neptune Aerocapture Entry Vehicle Preliminary Design" paper AIAA-2002-4812, AIAA Atmospheric Flight Mechanics Conference, Monterey, CA., Aug. 5-8, 2002.
6. R. Shotwell, T. R. Spilker, JPL, "Missions to Neptune and Triton in the Next Decade (2013-2023)", viewgraph presentation, Jan 24 2002.
7. R.W. Bailey, T.R. Spilker, R.J. Haw, JPL. Private communication and collaborative analysis, Dec. 2002 – Jan 2003.
8. G. Fraschetti, JPL, "MRO Optical Navigation Camera Preliminary Design Review", viewgraph presentation, May 8 2002.
9. W.M. Owen, JPL, private communications, April 2002 - May 2003.
10. R.J. Haw, "Approach Navigation for a Titan Aerocapture Orbiter", paper AIAA-2003-4802, AIAA Joint Propulsion Conference, Huntsville Alabama, July 20-23, 2003.
11. N. Strange, T. Goodson, Y. Hahn, "Cassini Tour Redesign for the Huygens Mission", paper AIAA-2002-4720, AIAA Specialist Conference, Monterey Calif, 5-8 August 2002.

APPENDIX 1: FLIGHT PATH ANGLE AND *B*-PLANE

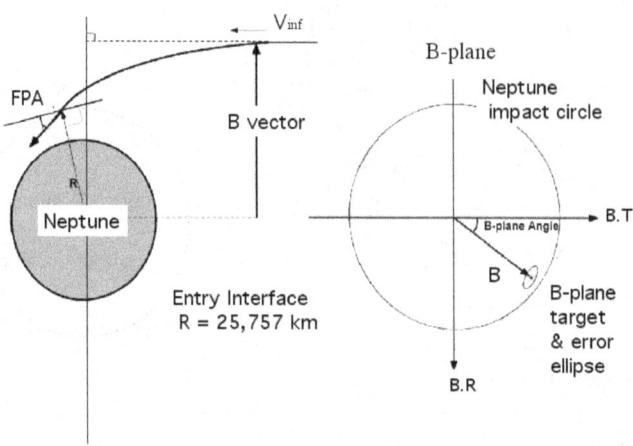

APPENDIX 2: *B*-PLANE DESCRIPTION

Planet or satellite approach trajectories are typically described in aiming plane coordinates referred to as "*B*–plane" coordinates (see Figure). The B-plane is a plane passing through the body center and perpendicular to the asymptote of the incoming trajectory (assuming two body conic motion). The "B-vector" is a vector in that plane, from the body center to the piercing-point of the trajectory asymptote. The B-vector specifies where the point of closest approach would be if the target body had no mass and did not deflect the flight path. Coordinate axes are defined by three orthogonal unit vectors, **S**, **T**, and **R**, with the system origin at the center of the target body. **S** is parallel to the spacecraft v_∞ vector (approximately the velocity vector at the time of entry into the target body's gravitational sphere of influence). **T** is arbitrary, but typically specified to lie in the ecliptic plane (the mean plane of the Earth's orbit), or in the body equatorial plane. Finally, **R** completes an orthogonal triad with **S** and **T**.

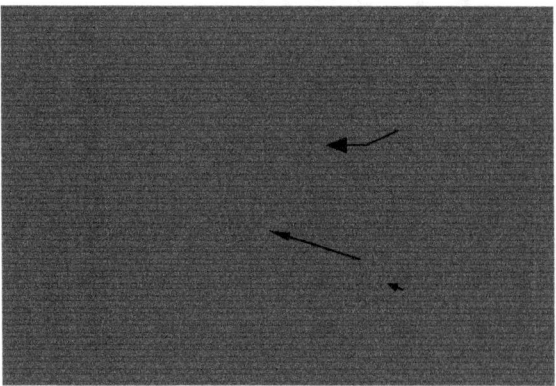

Aiming Plane Coordinate System Definition

Orbit determination errors can be characterized by a statistical dispersion ellipse in the aiming plane (*B*–plane) and a statistical uncertainty along the **S** (down-track) direction. In the Figure, SMIA and SMAA denote the semi–minor and semi–major axes of the dispersion ellipse (*i.e.* 50% of the distance across the ellipse along the respective coordinate). The angle θ is measured clockwise from **T** to SMAA.

APPENDIX 3: *A PRIORI* NAVIGATION MODEL UNCERTAINTY

Error Source	A Priori Uncertainty (1σ)	Correlation Time	Comments
Data			
doppler (mm/s)	0.05 / 0.075	Š	Ka-band (two-way / three-way doppler)
range (m)	5	Š	Ka-band
ΕDOR (nrad)	2	Š	0.055 ns (Ka-band)
optical (pixels) Triton only	0.25 - 0.5	Š	minimum stellar magnitude limit = 7.5
Estimated Parameters			
epoch state			
position (km)	1000	Š	
velocity (km/s)	1	Š	
Neptune ephemeris (km) $\times 10$	(3.4, 4., 1.7)	Š	R,AT,OOP (~ error at time of Voyager II)
doppler bias (mm/s)	0.0005	0	
range bias (m)	2	0	
clock bias (s)	1.0×10^8	0	
camera pointing error (deg)	(0.25, 0.25, 2)	0	R.A., Dec, Twist; estimated per observation
non-gravitational accelerations (km/s^2)	2.0×10^{12}	10 days	spherical covariance, estimated daily (1 day batches)
solar pressure	10%	Š	reflectivity coefficient
ACSΔV (mm/s), 1 per three weeks	(2, 2, 2)	Š	(line-of-sight, lateral, normal) components
TCMs (mm/s)			spherical covariance
TCM-1	5	Š	2% (3s) proportional error (per axis) 6 mm/s (3s) fixed error (per axis)
TCM-2	2	Š	10 milli-radian proportional pointing error (per axis)
TCM-3	15	Š	
Earth pole direction (cm)	$2 \rightarrow 5$	0	(X and Y). Ramps to higher value during final v of data.
UT1 (cm)	$2 \rightarrow 5$	0	(For UT1, ~5 cm -> 0.13 ms.)
ionosphere - day (cm)	55	0	S-band values
ionosphere - night (cm)	15	0	
troposphere (cm)	1	0	
Considered Parameters			
station locations (cm)	3	Š	
quasar locations (nrad)	2	Š	for ΔDOR data

ATMOSPHERIC MODELS FOR AEROCAPTURE

C. G. Justus and Aleta L. Duvall
NASA MSFC ED44/Morgan Research, Marshall Space Flight Center, AL, 35812

and

Vernon W. Keller
NASA MSFC ED44, Marshall Space Flight Center, AL, 35812

There are eight destinations in the Solar System with sufficient atmosphere for aerocapture to be a viable aeroassist option: Venus, Earth, Mars, Jupiter, Saturn and its moon Titan, Uranus, and Neptune. Engineering-level atmospheric models for four of these targets - Earth, Mars, Titan, and Neptune - have been developed for NASA to support systems analysis studies of potential future aerocapture missions. Development of a similar atmospheric model for Venus has recently commenced. An important capability of all of these models is their ability to simulate quasi-random density perturbations for Monte Carlo analysis in developing guidance, navigation and control algorithms, and for thermal systems design. Similarities and differences among these atmospheric models are presented, with emphasis on the recently developed Neptune model and on planned characteristics of the Venus model. Example applications for aerocapture are also presented and illustrated. Recent updates to the Titan atmospheric model are discussed, in anticipation of application to trajectory and atmospheric reconstruct for the Huygens Probe entry at Titan.

NOMENCLATURE

g	=	acceleration of gravity
H	=	atmospheric density scale height
Ls	=	planetocentric longitude of the Sun
M	=	mean molecular mass of atmospheric constituents
R	=	universal gas constant
T	=	atmospheric temperature

INTRODUCTION

ENGINEERING-LEVEL atmospheric models have been developed, or are under development, for five of the eight possible Solar System destinations where aerocapture could be used. These include Global Reference Atmospheric Models (GRAMs) for Earth (GRAM-99)[1,2], Mars (Mars-GRAM 2001)[3-6], Titan (Titan-GRAM)[7], Neptune (Neptune-GRAM)[8], and Venus-GRAM (under development). Physical characteristics of the various planetary atmospheres vary significantly. Likewise, significant variation is found in the amount of available data on which to base the respective engineering-level atmospheric models. The detailed characteristics of these models differ accordingly.

Earth-GRAM is based on climatology assembled from extensive observations by balloon, aircraft, ground-based remote sensing, sounding rockets, and satellite remote sensing. Details are provided in the GRAM User's Guide[1]. Mars-GRAM is based on climatologies of General Circulation Model (GCM) output, with details given in the Mars-GRAM User's Guide[3]. Mars-GRAM has been validated[4-6] by comparisons against observations made by Mars Global Surveyor, and against output from a separate Mars GCM. In contrast, data used to build Titan-GRAM and Neptune-GRAM are more limited, deriving primarily from Voyager observations and limited ground-based stellar occultation measurements. Titan-GRAM is based on data summarized in Ref. 9, while Neptune-GRAM was built from summaries of data contained in Ref. 10. For Venus, a substantial amount of data has been collected from orbiter and entry probe observations. These have been summarized in the Venus International Reference Atmosphere (VIRA)[11], which forms the basis for Venus-GRAM (under development).

Figure 1 shows the wide variety of temperature profiles encountered among the planets and Titan. For Earth, Venus, Mars, and Titan, height is measured from a reference surface (mean sea level on Earth). On Neptune, height is measured above the level at which pressure is one bar (Earth normal sea-level pressure). All of the planets exhibit a troposphere region, where temperature decreases with altitude, indicative of heat flow upward from the surface (on average). All of the planets exhibit a thermosphere region, where (on average) temperature increases with altitude, because of absorption of heat flux from the Sun as it penetrates into the atmosphere. All of the planets have stratospheres, where temperature decrease above the surface diminishes, and remains relatively constant until the base of the thermosphere (Earth being the exception to this, where the presence of ozone and resultant atmospheric heating produces a local temperature maximum in Earth's stratosphere-mesosphere region).

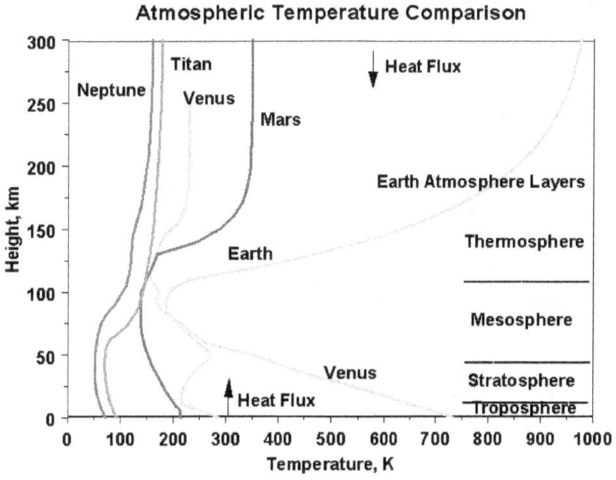

For interest in aerocapture or aerobraking, atmospheric density is the most important parameter. Fig. 2 compares density profiles on the planets and Titan. Vertical dashed lines in Fig. 2 indicate typical density values at which aerocapture or aerobraking operations would occur. Intersections of the aerocapture dashed line with various density curves shows that aerocapture would occur at a wide range of altitudes at the various destinations, varying from about 50 km at Mars to about 300 km at Titan. Aerobraking at Earth, Mars, and Venus would take place near, and just above, the 100 km level. At Neptune and Titan, aerobraking would be implemented near 550 km and 750 km, respectively.

Figure 2 shows that density decreases fairly

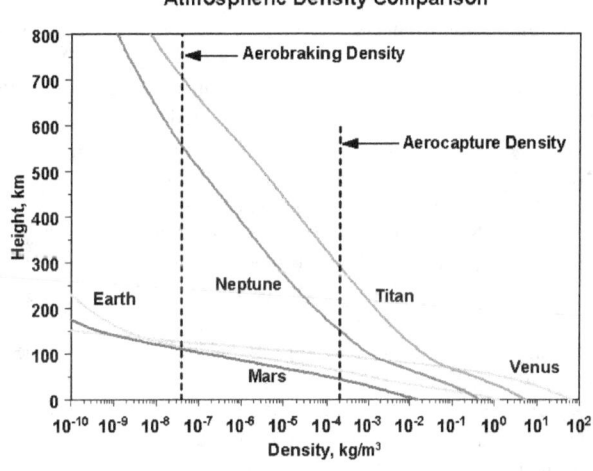

Figure 1: Comparison of temperature profiles among the planets and Titan.

Figure 2: Comparison of density profiles among the planets and Titan

rapidly with altitude for the terrestrial planets (Venus, Earth, Mars), while it decreases rather slowly for Neptune and Titan. This effect is explained by differences in density scale height, H, for the various planets and Titan. Density decreases rapidly with altitude if H is small, while it decreases slowly if H is large. H is proportional to pressure scale height [$R T / (M g)$]. For the terrestrial planets, molecular mass M is large ($M \approx 29$-44), so H is small. On Neptune, H is large because M is small for Neptune's hydrogen-helium atmosphere ($M \approx 2$). For Titan, H is large despite the high molecular mass of its atmosphere (M $\approx$ 29), because its gravity is low.

BASIS FOR THE ATMOSPHERIC MODELS

In Earth-GRAM, Mars-GRAM, and Venus-GRAM, input values for date, time, latitude, longitude, etc. are used to calculate planetary position and solar position. In this manner, effects of latitude variation and seasonal and time-of-day variations can be computed explicitly. A simplified approach is adopted in Titan-GRAM and Neptune-GRAM, whereby these effects, as well as effects of relatively large measurement uncertainties for these planets, are represented within a prescribed envelope of minimum-average-maximum density versus altitude. Figure 3 shows this envelope for Titan. Engineering atmospheric model data developed for the Huygens entry probe[9] are used to define the Titan envelope. For Neptune, data from Ref. 10 are employed to generate a comparable minimum-maximum envelope, as shown in Fig. 4.

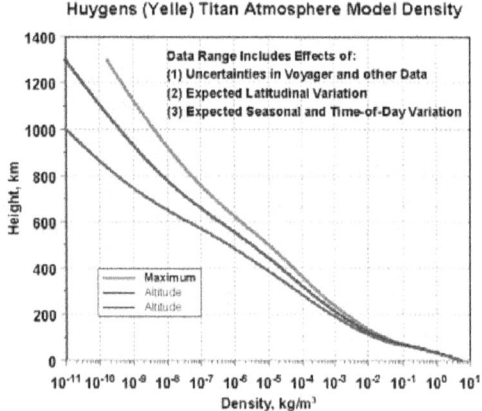

Figure 3: Minimum, average, and maximum density profiles for Titan[9]

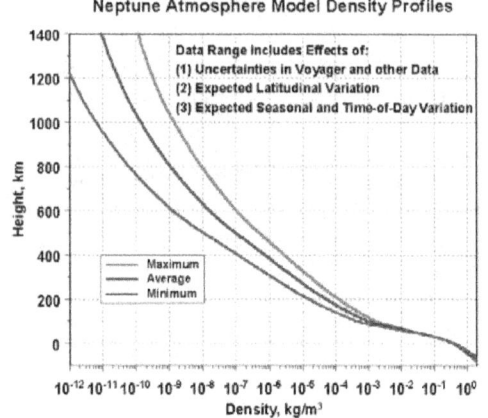

Figure 4: Minimum, average, and maximum density profiles for Neptune from data in Ref. 10.

A single model input parameter (Fminmax) allows the user of Titan-GRAM or Neptune-GRAM to select where within the min-max envelope a particular simulation will fall. Fminmax = -1, 0, or 1 selects minimum, average, or maximum conditions, respectively, with intermediate values determined by interpolation (i.e., Fminmax between 0 and 1 produces values between average and maximum). Effects such as variation with latitude along a given trajectory path can be computed using the appropriate representation of Fminmax variation with latitude.

Since drag is proportional to density, density is the most important atmospheric parameter for aerocapture. Next most important is height variation of density, as characterized by density scale height. Density scale height is important in determining aerocapture corridor width, or entry angle range that allows the vehicle to achieve capture orbit without "skipping out" or "burning in". As discussed above, small density scale height means rapid change of density with altitude, which results in low corridor width. Large density scale height implies slow density change with altitude, and large corridor width.

Figure 5 compares height profiles of density scale height among the planets and Titan. Aerocapture altitude (c.f. discussion of Fig 2) is indicated by letter A in Fig. 5. This figure shows low density scale heights (4-8 km) at aerocapture altitudes for the terrestrial planets. Larger scale heights (≈ 30-50 km) occur at aerocapture altitudes on Neptune and Titan.

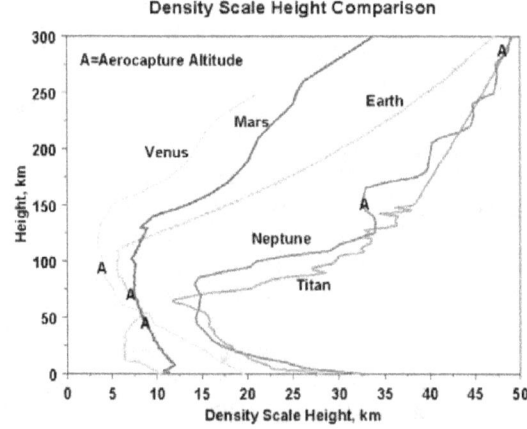

Figure 5: Comparison of atmospheric density scale height among the planets and Titan

TITAN-GRAM GCM OPTION

An option has recently been added for using Titan General Circulation Model (GCM) data as input for Titan-GRAM. These Titan GCM data are derived from graphs in Ref 12. Upper altitudes for the Titan GCM option are computed using a parameterized fit to Titan exospheric temperatures, taken from graphs in Ref 13. Figure 6 shows a height-latitude cross section of density, expressed as percent deviation from the mean, for Voyager encounter date

November 12, 1980 (planetocentric longitude of Sun $Ls = 8.8°$), 00:00 GMT, longitude zero, local solar time 0.7 Titan hours. Figure 7 compares vertical density profiles at latitude zero, local solar time 1 hour and 13 hours on the Voyager encounter date, with the Huygens Yelle[9] minimum-maximum density envelope from Fig. 3. This figure shows that the Titan GCM results correspond fairly closely with Yelle maximum conditions up to about 300 km altitude, and agree quite closely with Yelle average conditions (vertical line at 0 in Fig. 7) above about 500 km.

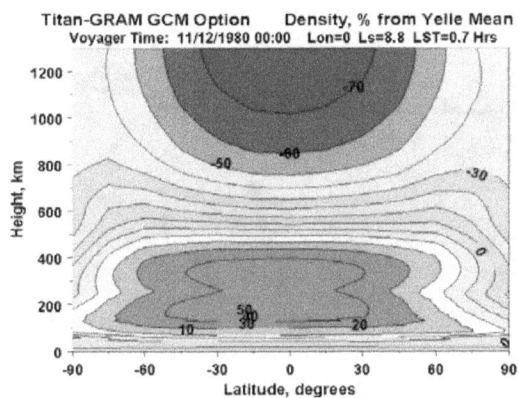

Figure 6: Density (percent deviation from mean) versus height and latitude, using Titan-GRAM GCM option.

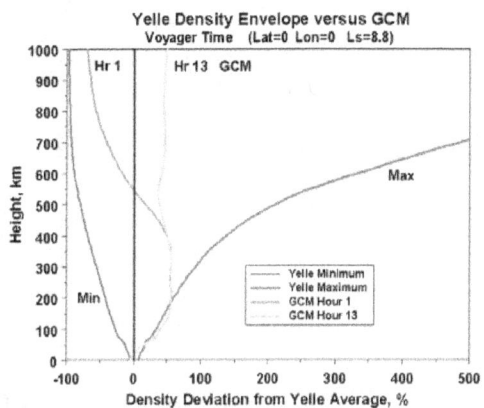

Figure 7: Comparison of two selected Titan-GRAM density profiles (GCM option) with minimum-maximum envelope from Huygens Yelle model[9].

VENUS-GRAM DEVELOPMENT

Based on the Venus International Reference Atmosphere (VIRA)[11], Venus-GRAM is being developed and applied in ongoing Venus aerocapture performance analyses. Figure 8 shows a plot of density (percent deviation from the mean) versus height and latitude from Venus-GRAM. Conditions in Fig. 8 are for $Ls = 90°$ and local solar time = 12 Venus hours.

Below about 100 km altitude on Venus, we find that temperature, density, and density scale height conditions are very uniform with both latitude and time of day. VIRA data below 100 km altitude vary only slightly with latitude and have no dependence on local solar time. Between 100 km and 150 km, VIRA data depend on local solar time, but not latitude. From 150 km to its top at 250 km, VIRA depends on solar zenith angle, which is affected by both latitude and local solar time.

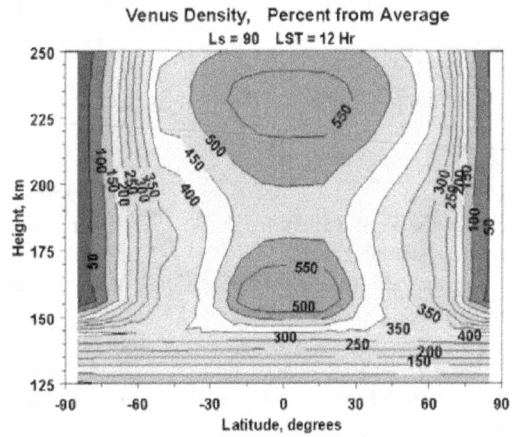

Figure 8: Example height-latitude density cross section from Venus-GRAM.

PERTURBATION MODELS

An important feature of all the GRAM atmospheric models is their ability to simulate "high frequency" perturbations in density and winds, due to such phenomena as turbulence and various kinds of atmospheric waves. As illustrated in Fig. 9, Earth-GRAM altitude, latitude, and monthly variations of perturbation standard deviations are based on a large climatology of observations. For Titan-GRAM and Neptune-GRAM, perturbation standard deviations are computed from an analytical expression for gravity wave saturation conditions, explained more fully in Ref. 7. As shown in Fig. 9, the resulting vertical profiles of standard deviations for Titan and Neptune are not dissimilar to Earth observations, when expressed as percent of mean density. For Mars-GRAM, a similar gravity wave saturation relation is used to estimate density perturbation standard deviations, except that effects of significant topographic variation on Mars are also taken into account. Up to about 75 km altitude, the Mars model density standard deviations are also fairly consistent with Earth observations. By about 100 km to 130 km altitude, Mars model density standard deviations increase to about 20% to 35% of mean value, consistent with observed orbit-to-orbit density variations observed by Mars Global Surveyor and Mars Odyssey.

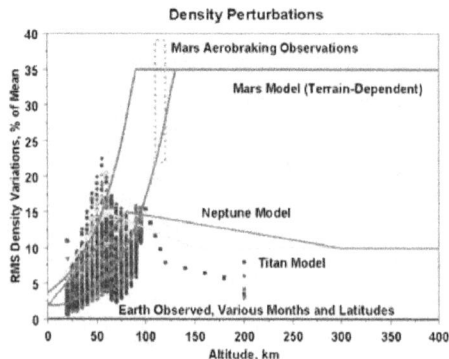

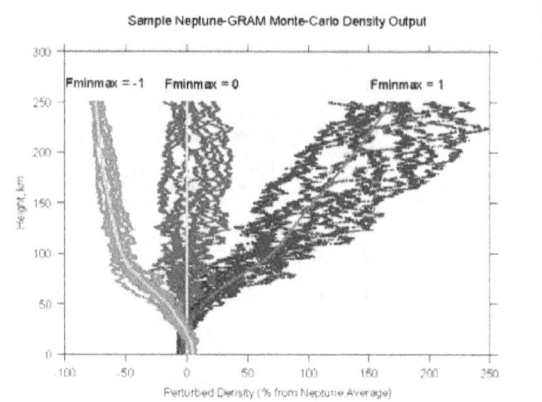

Figure 9: Height variation of density perturbation model standard deviations for Earth, Mars, Titan, and Neptune.

Figure 10: Sample Monte Carlo density perturbations from Neptune-GRAM, expressed as percent deviation from Neptune mean value.

A typical application of the Neptune-GRAM perturbation model is shown in Figure 10. Neptune-GRAM was recently utilized in Neptune aerocapture systems studies for trajectory analysis. The chosen aerocapture design reference mission included simulations which involved capture into a highly eccentric orbit, to allow the orbiter to periodically visit Triton for scientific observations. The ability to successfully aerocapture into such an eccentric orbit depends very significantly on details of Monte Carlo trajectory simulations, particularly on atmospheric density variations such as illustrated in Fig 10. For such an eccentric orbit, there is relatively little margin for error between a captured orbit and one which exceeds escape velocity upon atmospheric exit, a result which could ultimately lead to mission failure. Neptune-GRAM was used to define an aerocapture corridor width consistent with mission success.

CONCLUSIONS

The engineering-level atmospheric models presented here are suitable for a wide range of mission design, systems analysis, and operations tasks. For orbiter missions, applications include analysis for aerocapture or aerobraking operations, analysis of station-keeping issues for science orbits, analysis of orbital lifetimes for end-of-mission planetary protection orbits, and atmospheric entry issues for accidental break-up and burn-up scenarios. For lander missions to Venus, Mars and Titan, and for Earth-return, applications include analysis for entry, descent and landing (EDL), and guidance, navigation and control analysis for precision landing and hazard avoidance. Perturbation simulation capabilities of these models make them especially useful in Monte Carlo analyses for design and testing of guidance, navigation, and control algorithms, and for heat loads analysis of thermal protection systems.

ACKNOWLEDGMENTS

The authors gratefully acknowledge support from the NASA Marshall Space Flight Center In-Space Propulsion Program. Particular thanks go to Bonnie James (MSFC), Manager of the Aerocapture Technology Development Project, to Michelle M. Munk (LaRC/MSFC), Lead Systems Engineer for Aerocapture, and to Melody Herrmann (MSFC), team lead and Mary Kae Lockwood (LaRC), technical lead for the Titan/Neptune Systems Analysis study. Model user feedback and suggestions from the following individuals are also greatly appreciated: Dick Powell, Brett Starr, and David Way (NASA LaRC), and Claude Graves, Jim Masciarelli, Lee Bryant, Tim Crull, and Tom Smith (NASA JSC). External review comments from Prof. Darrell Strobel (Johns Hopkins University) were especially helpful.

REFERENCES

[1]Justus, C. G., and Johnson, D. L., "The NASA/MSFC Global Reference Atmospheric Model - 1999 Version (GRAM-99)", NASA/TM-1999-209630, 1999.

[2]Justus, C.G., Duvall, A. L., and Johnson, D. L., "Earth Global Reference Atmospheric Model and Trace Constituents", *34th COSPAR Scientific Assembly*, Houston, Texas, Invited Paper C4.1-0005-02, October, 2002.

[3]Justus, C. G., and Johnson, D. L., "Mars Global Reference Atmospheric Model 2001 Version (Mars-GRAM 2001) Users Guide", NASA/TM-2001-210961, April, 2001.

[4]Justus, C. G., Duvall, A. L., and Johnson, D. L., "Mars-GRAM Validation with Mars Global Surveyor Data", *34th COSPAR Scientific Assembly*, Houston, Texas, Paper C3.3-0029-02, October, 2002.

[5]Justus, C. G., Duvall, A. L., and Johnson, D. L., "Global MGS TES Data and Mars-GRAM Validation", *34th COSPAR Scientific Assembly*, Houston, Texas, Paper C4.2-0005-02, October, 2002.

[6]Justus, C. G., Duvall, A. L. and Johnson, D. L., "Mars Global Reference Atmospheric Model (Mars-GRAM) and Database for Mission Design", *International Workshop on Mars Atmosphere Modeling and Observations*, Granada, Spain, January, 2003.

[7]Justus, C.G., Duvall, A. L., and Johnson, D. L., "Engineering-level model atmospheres for Titan and Neptune", *39th AIAA/ASME/SAE/ASEE Joint Propulsion Conference*, Huntsville, Alabama, Paper AIAA-2003-4803, July, 2003.

[8]Justus, C.G., Duvall, A. L., and Keller, V. W., "Engineering-level model atmospheres for Titan and Mars", *International Workshop on Planetary Probe Atmospheric Entry and Descent Trajectory Analysis and Science*, Lisbon, Portugal. October, 2003.

[9]Yelle, R.V. , Strobell, D. F., Lellouch, E., and Gautier, D., " Engineering Models for Titan's Atmosphere", in *Huygens Science, Payload and Mission*, ESA SP-1177, August, 1997.

[10]Cruikshank, D.P. (ed.), *Neptune and Triton*, University of Arizona Press, Tucson, 1995.

[11]Kliore, A. J., Moroz, V. I., and Keating, G. M. (eds.), "The Venus International Reference Atmosphere", *Advances in Space Research*, vol. 5, no. 11, 1985, Pergamon Press, Oxford, 1986, pp. 1-304.

[12]Hourdin, F., Talagrand, O., Sadourny, R., Courtin, R., Gautier, D., and McKay, C.P., "Numerical simulation of the general circulation of the atmosphere of Titan", *Icarus*, vol. 117, no. 2, Oct. 1995, pp. 358-74.

[13]Mueller-Wodarg, I. C. F., "The Application of General Circulation Models to the Atmospheres of Terrestrial-Type Moons of the Giant Planets", in *Comparative Atmospheres in the Solar System*, American Geophysical Union, 2002.

ATMOSPHERIC MODELS FOR AEROCAPTURE SYSTEMS STUDIES

C. G. Justus and Aleta L. Duvall
NASA MSFC ED44/Morgan Research, Marshall Space Flight Center, AL, 35812

and

Vernon W. Keller
NASA MSFC ED44, Marshall Space Flight Center, AL, 35812

Aerocapture uses atmospheric drag to decelerate into captured orbit from interplanetary transfer orbit. This includes capture into Earth orbit from, for example, Lunar-return or Mars-return orbit. Eight Solar System destinations have sufficient atmosphere for aerocapture to be applicable – three of the rocky planets (Venus, Earth, and Mars), four gas giants (Jupiter, Saturn, Uranus, and Neptune), and Saturn's moon Titan. These destinations fall into two general groups: (1) The rocky planets, which have warm surface temperatures (about 200 K to 750 K) and rapid decrease of density with altitude, and (2) the gas giants and Titan, which have cold temperatures (about 70 K to 170 K) at the surface or 1-bar pressure level, and slow rate of decrease of density with altitude. Aerocapture altitudes at the gas giants typically range from about 150 km to 300 km above the 1-bar pressure reference. Aerocapture at the rocky planets would occur at altitudes of about 50 km to 100 km. In contrast, aerobraking (circularizing a highly elliptical capture orbit, using multiple atmospheric passes) would occur at widely varying altitudes ranging from about 125 km out to 700 km for Titan. In addition to aerocapture altitude, aerocapture corridor width is also determined by details of the atmospheric density profile. Corridor width is the range of atmospheric entry angles allowable for successful aerocapture, i.e., achieving capture orbit without "skip-out" or "burn-up". Corridor width is significantly affected by rate of change of density, as measured by the density scale height, at aerocapture periapsis altitude. Density scale height is the vertical distance over which density changes by a factor of e, (i.e., logarithmically). Larger scale height values mean slower density variation with height and larger corridor width; smaller scale height leads to smaller corridor width. For the rocky planets, the overall rapid fall-off of density with height leads to relatively low density scale heights at aerocapture altitudes and small aerocapture corridor widths for these destinations. Larger density scale heights, with consequently larger corridor widths, result from the slower density fall-of with height for Titan and the gas giant planets. Density scale height values at periapsis for the rocky planets vary from about 4 km to 8 km; for the gas giant planets and Titan this range is about 25 km to 50 km. Engineering-level atmospheric models for Earth, Mars, Titan, and Neptune have been developed for NASA systems analysis studies of potential future aerocapture missions. Development of a similar atmospheric model for Venus has recently commenced. These models are collectively referred to as Global Reference Atmosphere Models, or GRAMs. An important capability of all of the GRAM models is their ability to simulate quasi-random density perturbations for Monte Carlo analyses in developing guidance, navigation, and control algorithms, and for thermal systems design. Small-scale root-mean-square (rms) density perturbations observed for Earth may be compared with those modeled for Mars, Titan, and Neptune. Monte-Carlo simulations of density variations for Neptune atmospheric conditions yield minimum, average, and maximum density profiles due to expected variations with season, latitude, time-of-day, etc. Details of these comparisons and simulations are discussed.

NOMENCLATURE

g	=	acceleration of gravity
H	=	atmospheric density scale height
Ls	=	planetocentric longitude of the Sun
M	=	mean molecular mass of atmospheric constituents
R	=	universal gas constant
T	=	atmospheric temperature

INTRODUCTION

ENGINEERING-LEVEL atmospheric models have been developed, or are under development, for five of the eight possible Solar System destinations where aerocapture could be used. These include Global Reference Atmospheric Models (GRAMs) for Earth (GRAM-99)[1,2], Mars (Mars-GRAM 2001)[3-6], Titan (Titan-GRAM)[7], Neptune (Neptune-GRAM)[8], and Venus-GRAM (under development). Physical characteristics of the various planetary atmospheres vary significantly. Likewise, significant variation is found in the amount of available data on which to base the respective engineering-level atmospheric models. The detailed characteristics of these models differ accordingly.

Earth-GRAM is based on climatology assembled from extensive observations by balloon, aircraft, ground-based remote sensing, sounding rockets, and satellite remote sensing. Details are provided in the GRAM User's Guide[1]. Mars-GRAM is based on climatologies of General Circulation Model (GCM) output, with details given in the Mars-GRAM User's Guide[3]. Mars-GRAM has been validated[4-6] by comparisons against observations made by Mars Global Surveyor, and against output from a separate Mars GCM. In contrast, data used to build Titan-GRAM and Neptune-GRAM are more limited, deriving primarily from Voyager observations and limited ground-based stellar occultation measurements. Titan-GRAM is based on data summarized in Ref. 9, while Neptune-GRAM was built from summaries of data contained in Ref. 10. For Venus, a substantial amount of data has been collected from orbiter and entry probe observations. These have been summarized in the Venus International Reference Atmosphere (VIRA)[11], which forms the basis for Venus-GRAM (under development).

Figure 1 shows the wide variety of temperature profiles encountered among the planets and Titan. For Earth, Venus, Mars, and Titan, height is measured from a reference surface (mean sea level on Earth). On Neptune, height is measured above the level at which pressure is one bar (Earth normal sea-level pressure). All of the planets exhibit a troposphere region, where temperature decreases with altitude, indicative of heat flow upward from the surface (on average). All of the planets exhibit a thermosphere region, where (on average) temperature increases with altitude, because of absorption of heat flux from the Sun as it penetrates into the atmosphere. All of the planets have stratospheres, where temperature decrease above the surface diminishes, and remains relatively constant until the base of the thermosphere (Earth being the exception to this, where the presence of ozone and resultant atmospheric heating produces a local temperature maximum in Earth's stratosphere-mesosphere region).

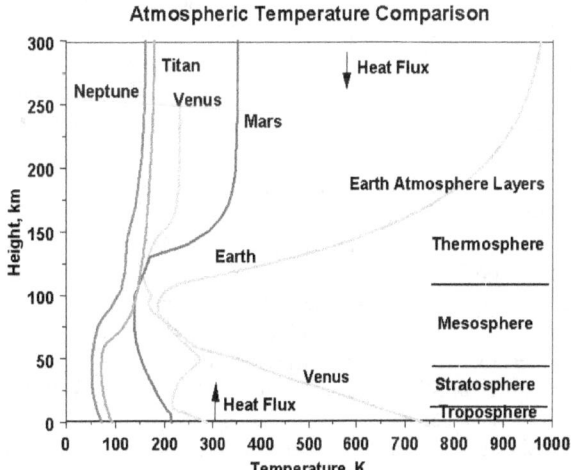

Figure 1: Comparison of temperature profiles among the planets and Titan.

For interest in aerocapture or aerobraking, atmospheric density is the most important parameter. Figures 2 and 3 compare density profiles for the gas giants and the rocky planets. Vertical dashed lines in the figures indicate typical density values at which aerocapture or aerobraking operations would occur. Intersections of the aerocapture dashed line with various density curves shows that aerocapture would occur at a wide range of altitudes at the various destinations, varying from about 50 km at Mars to about 300 km at Titan. Aerobraking at Earth, Mars, and Venus would take place near, and just above, the 100 km level. At Neptune and Titan, aerobraking would be implemented near 550 km and 750 km, respectively.

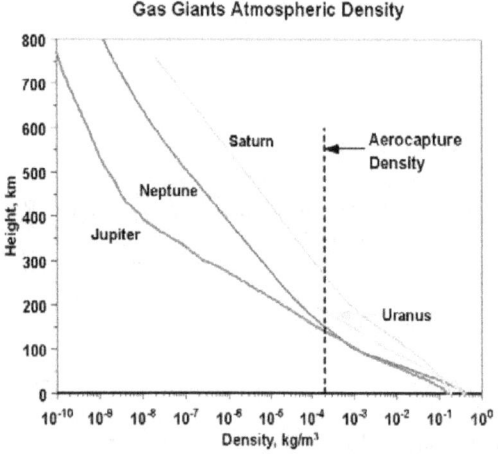

Figure 2: Comparison of atmospheric density profiles for the gas giant planets

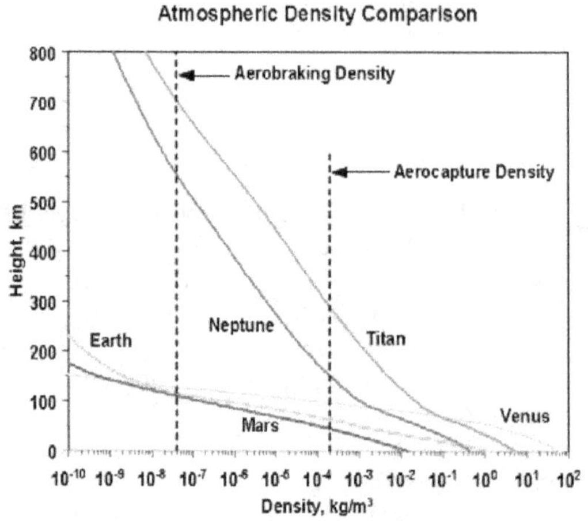

Figure 3: Density profiles for rocky planets compared with those for Titan and Neptune

Figures 2 and 3 show that density decreases fairly rapidly with altitude for the terrestrial planets (Venus, Earth, Mars), while it decreases rather slowly for the gas giants and Titan. This effect is explained by differences in density scale height, H, for the various planets and Titan. Density decreases rapidly with altitude if H is small, while it decreases slowly if H is large. H is proportional to pressure scale height $[R T / (M g)]$. For the terrestrial planets, molecular mass M is large ($M \approx$ 29-44), so H is small. For the gas giants, H is large because M is small ($M \approx$ 2.1-2.7) for predominantly hydrogen-helium atmospheres. For Titan, H is large despite the high molecular mass of its atmosphere (M $\approx$ 29), because its gravity is low.

BASIS FOR THE ATMOSPHERIC MODELS

In Earth-GRAM, Mars-GRAM, and Venus-GRAM, input values for date, time, latitude, longitude, etc. are used to calculate planetary position and solar position. In this manner, effects of latitude variation and seasonal and time-of-day variations can be computed explicitly. A simplified approach is adopted in Titan-GRAM and Neptune-GRAM, whereby these effects, as well as effects of relatively large measurement uncertainties for these planets, are represented within a prescribed envelope of minimum-average-maximum density versus altitude. Figure 4 shows this envelope for Titan. Engineering atmospheric model data developed for the Huygens entry probe[9] are used to define the Titan envelope. For Neptune, data from Ref. 10 are employed to generate a comparable minimum-maximum envelope, as shown in Fig. 5.

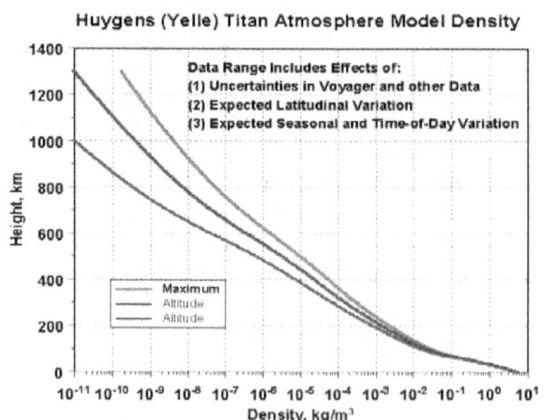

Figure 4: Minimum, average, and maximum density

A single model input parameter (Fminmax) allows the user of Titan-GRAM or Neptune-GRAM to select where within the min-max envelope a particular simulation will fall. Fminmax = -1, 0, or 1 selects minimum, average, or maximum conditions, respectively, with intermediate values determined by interpolation (i.e., Fminmax between 0 and 1 produces values between average and maximum). Effects such as variation with latitude along a given trajectory path can be computed using the appropriate representation of Fminmax variation with latitude. Since drag is proportional to density, density is the most important atmospheric parameter for aerocapture.

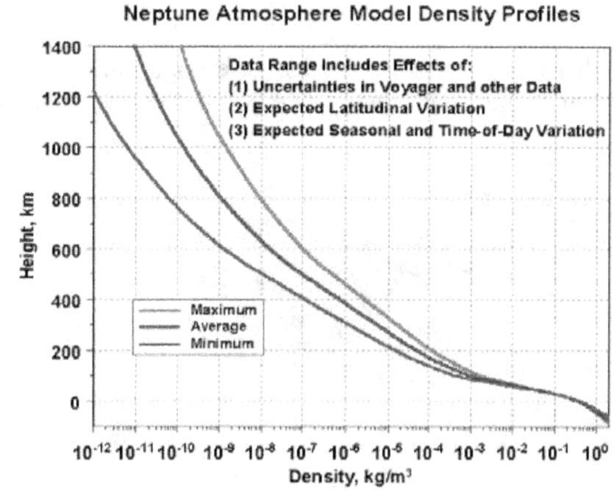

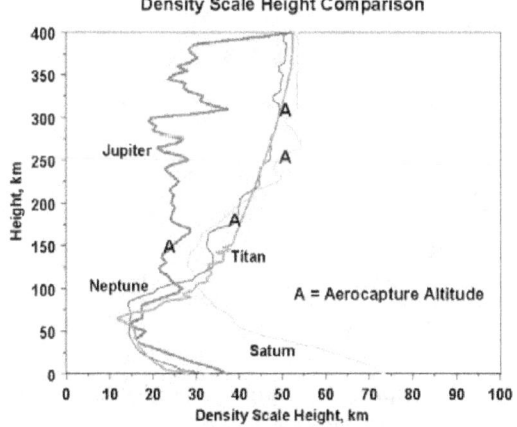

Figure 5: Minimum, average, and maximum density profiles for Neptune from data in Ref. 10.

Next most important is height variation of density, as characterized by density scale height. Density scale height is important in determining aerocapture corridor width, or entry angle range that allows the vehicle to achieve capture orbit without "skipping out" or "burning in". As discussed above, small density scale height means rapid change of density with altitude, which results in low corridor width. Large density scale height implies slow density change with altitude, and large corridor width.

Figures 6 and 7 compare height profiles of density scale height among the gas giants and the rocky planets. Aerocapture altitude (c.f. discussion of Fig 2) is indicated by letter A in Figs. 6 and 7. These figures show low density scale heights (4-8 km) at aerocapture altitudes for the terrestrial planets. Larger scale heights ($\approx$ 30-50 km) occur at aerocapture altitudes on Neptune and Titan.

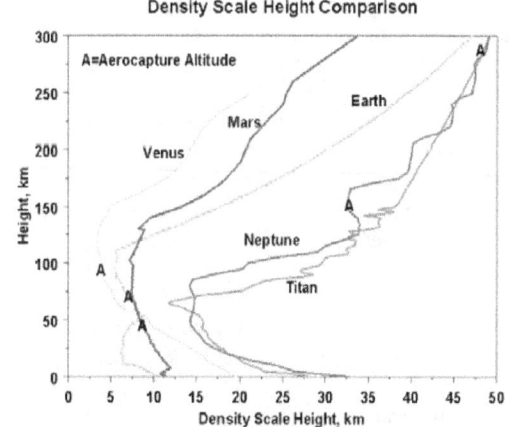

Figure 6: Density scale height profiles for Jupiter, Saturn, Neptune, and Titan

Figure 7: Density scale height profiles for the rocky planets compared to those for Titan and Neptune

TITAN-GRAM GCM OPTION

An option has recently been added for using Titan General Circulation Model (GCM) data as input for Titan-GRAM. These Titan GCM data are derived from graphs in Ref 12. Upper altitudes for the Titan GCM option are computed using a parameterized fit to Titan exospheric temperatures, taken from graphs in Ref 13. Figure 8 shows a height-latitude cross section of density, expressed as percent deviation from the mean, for Voyager encounter date November 12, 1980 (planetocentric longitude of Sun Ls = 8.8°), 00:00 GMT, longitude zero, local solar time 0.7 Titan hours. Figure 9 compares vertical density profiles at latitude zero, local solar time 1 hour and 13 hours on the Voyager encounter date, with the Huygens Yelle[9] minimum-maximum density envelope from Fig. 4. This figure shows that the Titan GCM results correspond fairly closely with Yelle maximum conditions up to about 300 km altitude, and agree quite closely with Yelle average conditions (vertical line at 0 in Fig. 9) above about 500 km.

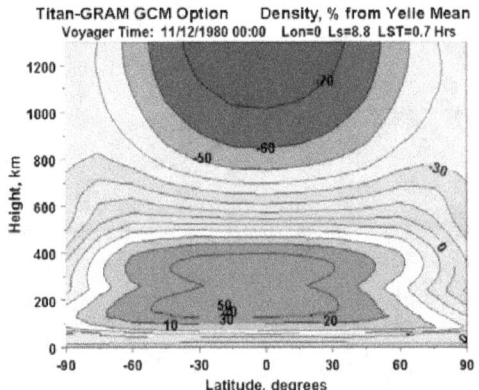

Figure 8: Density (percent deviation from mean) versus height and latitude, using Titan-GRAM GCM option.

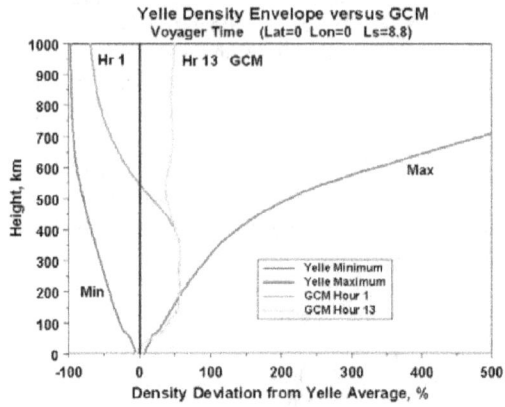

Figure 9: Comparison of two selected Titan-GRAM density profiles (GCM option) with minimum-maximum envelope from Huygens Yelle model[9].

VENUS-GRAM DEVELOPMENT

Based on the Venus International Reference Atmosphere (VIRA)[11], Venus-GRAM is being developed and applied in ongoing Venus aerocapture performance analyses. Figure 10 shows a plot of density (percent deviation from the mean) versus height and latitude from Venus-GRAM. Conditions in Fig. 10 are for $Ls = 90°$ and local solar time = 12 Venus hours.

Below about 100 km altitude on Venus, we find that temperature, density, and density scale height conditions are very uniform with both latitude and time of day. VIRA data below 100 km altitude vary only slightly with latitude and have no dependence on local solar time. Between 100 km and 150 km, VIRA data depend on local solar time, but not latitude. From 150 km to its top at 250 km, VIRA depends on solar zenith angle, which is affected by both latitude and local solar time.

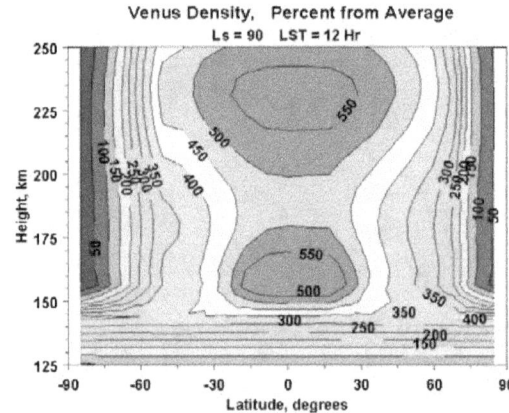

Figure 10: Example height-latitude density cross section from Venus-GRAM.

PERTURBATION MODELS

An important feature of all the GRAM atmospheric models is their ability to simulate "high frequency" perturbations in density and winds, due to such phenomena as turbulence and various kinds of atmospheric waves. As illustrated in Fig. 11, Earth-GRAM altitude, latitude, and monthly variations of perturbation standard deviations are based on a large climatology of observations. For Titan-GRAM and Neptune-GRAM, perturbation standard deviations are computed from an analytical expression for gravity wave saturation conditions, explained more fully in Ref. 7. As shown in Fig. 11, the resulting vertical profiles of standard deviations for Titan and Neptune are not dissimilar to Earth observations, when expressed as percent of mean density. For Mars-GRAM, a similar gravity wave saturation relation is used to estimate density perturbation standard deviations, except that effects of significant topographic variation on Mars are also taken into account. Up to about 75 km altitude, the Mars model density standard deviations are also fairly consistent with Earth observations. By about 100 km to 130 km altitude, Mars model density standard deviations increase to about 20% to 35% of mean value, consistent with observed orbit-to-orbit density variations observed by Mars Global Surveyor and Mars Odyssey.

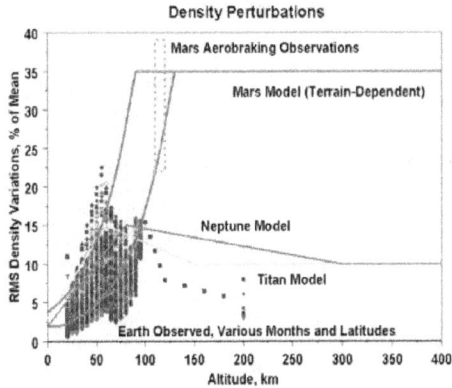

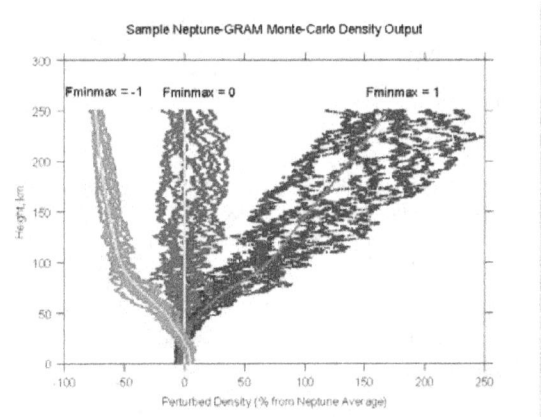

Figure 11: Height variation of density perturbation model standard deviations for Earth, Mars, Titan, and Neptune.

Figure 12: Sample Monte Carlo density perturbations from Neptune-GRAM, expressed as percent deviation from Neptune mean value.

A typical application of the Neptune-GRAM perturbation model is shown in Fig. 12. Neptune-GRAM was recently utilized in Neptune aerocapture systems studies for trajectory analysis. The chosen aerocapture design reference mission included simulations which involved capture into a highly eccentric orbit, to allow the orbiter to periodically visit Triton for scientific observations. The ability to successfully aerocapture into such an eccentric orbit depends very significantly on details of Monte Carlo trajectory simulations, particularly on atmospheric density variations such as illustrated in Fig. 12. For such an eccentric orbit, there is relatively little margin for error between a captured orbit and one which exceeds escape velocity upon atmospheric exit, a result which could ultimately lead to mission failure. Neptune-GRAM was used to define an aerocapture corridor width consistent with mission success.

CONCLUSIONS

The engineering-level atmospheric models presented here are suitable for a wide range of mission design, systems analysis, and operations tasks. For orbiter missions, applications include analysis for aerocapture or aerobraking operations, analysis of station-keeping issues for science orbits, analysis of orbital lifetimes for end-of-mission planetary protection orbits, and atmospheric entry issues for accidental break-up and burn-up scenarios. For lander missions to Venus, Mars and Titan, and for Earth-return, applications include analysis for entry, descent and landing (EDL), and guidance, navigation and control analysis for precision landing and hazard avoidance. Perturbation simulation capabilities of these models make them especially useful in Monte Carlo analyses for design and testing of guidance, navigation, and control algorithms, and for heat loads analysis of thermal protection systems.

ACKNOWLEDGMENTS

The authors gratefully acknowledge support from the NASA Marshall Space Flight Center In-Space Propulsion Program. Particular thanks go to Bonnie James (MSFC), Manager of the Aerocapture Technology Development Project, to Michelle M. Munk (LaRC/MSFC), Lead Systems Engineer for Aerocapture, and to Melody Herrmann (MSFC), team lead and Mary Kae Lockwood (LaRC), technical lead for the Titan/Neptune Systems Analysis study. Model user feedback and suggestions from the following individuals are also greatly appreciated: Dick Powell, Brett Starr, and David Way (NASA LaRC), and Claude Graves, Jim Masciarelli, Lee Bryant, Tim Crull, and Tom Smith (NASA JSC). External review comments from Prof. Darrell Strobel (Johns Hopkins University) were especially helpful.

REFERENCES

[1]Justus, C. G., and Johnson, D. L., "The NASA/MSFC Global Reference Atmospheric Model - 1999 Version (GRAM-99)," NASA/TM-1999-209630, 1999.

[2]Justus, C.G., Duvall, A. L., and Johnson, D. L., "Earth Global Reference Atmospheric Model and Trace Constituents," *Advances in Space Research*, in press.

[3]Justus, C. G., and Johnson, D. L., "Mars Global Reference Atmospheric Model 2001 Version (Mars-GRAM 2001) Users Guide," NASA/TM-2001-210961, April, 2001.

[4]Justus, C. G., Duvall, A. L., and Johnson, D. L., "Mars-GRAM Validation with Mars Global Surveyor Data," *34th COSPAR Scientific Assembly*, Houston, Texas, Paper C3.3-0029-02, October, 2002.

[5]Justus, C. G., Duvall, A. L., and Johnson, D. L., "Global MGS TES Data and Mars-GRAM Validation," *Advances in Space Research*, in press.

[6]Justus, C. G., Duvall, A. L. and Johnson, D. L., "Mars Global Reference Atmospheric Model (Mars-GRAM) and Database for Mission Design," *International Workshop on Mars Atmosphere Modeling and Observations*, Granada, Spain, January, 2003.

[7]Justus, C.G., Duvall, A. L., and Johnson, D. L., "Engineering-level model atmospheres for Titan and Neptune," AIAA-2003-4803, *39th AIAA/ASME/SAE/ASEE Joint Propulsion Conference*, Huntsville, Alabama, July 20-23, 2003.

[8]Justus, C.G., Duvall, A. L., and Keller, V. W., "Engineering-level model atmospheres for Titan and Mars," *International Workshop on Planetary Probe Atmospheric Entry and Descent Trajectory Analysis and Science*, Lisbon, Portugal. October, 2003.

[9]Yelle, R.V. , Strobell, D. F., Lellouch, E., and Gautier, D., " Engineering Models for Titan's Atmosphere," *Huygens Science, Payload and Mission*, ESA SP-1177, August, 1997.

[10]Cruikshank, D.P. (ed.), *Neptune and Triton*, University of Arizona Press, Tucson, 1995.

[11]Kliore, A. J., Moroz, V. I., and Keating, G. M. (eds.), "The Venus International Reference Atmosphere," *Advances in Space Research*, vol. 5, no. 11, 1985, Pergamon Press, Oxford, 1986, pp. 1-304.

[12]Hourdin, F., Talagrand, O., Sadourny, R., Courtin, R., Gautier, D., and McKay, C.P., "Numerical simulation of the general circulation of the atmosphere of Titan," *Icarus*, vol. 117, no. 2, Oct. 1995, pp. 358-74.

[13]Mueller-Wodarg, I. C. F., "The Application of General Circulation Models to the Atmospheres of Terrestrial-Type Moons of the Giant Planets," in *Comparative Atmospheres in the Solar System*, American Geophysical Union, 2002.

AEROCAPTURE PERFORMANCE ANALYSIS FOR A NEPTUNE-TRITON EXPLORATION MISSION

Brett R. Starr

NASA Langley Research Center, Hampton, Virginia, 23681-2199

Carlos H. Westhelle and James P. Masciarelli

NASA Johnson Space Flight Center, Houston, Texas, 77058

A systems analysis has been conducted for a Neptune-Triton Exploration Mission in which aerocapture is used to capture a spacecraft at Neptune. Aerocapture uses aerodynamic drag instead of propulsion to decelerate from the interplanetary approach trajectory to a captured orbit during a single pass through the atmosphere. After capture, propulsion is used to move the spacecraft from the initial captured orbit to the desired science orbit. A preliminary assessment identified that a spacecraft with a lift to drag ratio of 0.8 was required for aerocapture. Performance analyses of the 0.8 L/D vehicle were performed using a high fidelity flight simulation within a Monte Carlo executive to determine mission success statistics. The simulation was the Program to Optimize Simulated Trajectories (POST) modified to include Neptune specific atmospheric and planet models, spacecraft aerodynamic characteristics, and interplanetary trajectory models. To these were added autonomous guidance and pseudo flight controller models. The Monte Carlo analyses incorporated approach trajectory delivery errors, aerodynamic characteristics uncertainties, and atmospheric density variations. Monte Carlo analyses were performed for a reference set of uncertainties and sets of uncertainties modified to produce increased and reduced atmospheric variability. For the reference uncertainties, the 0.8 L/D flatbottom ellipsled vehicle achieves 100% successful capture and has a 99.87 probability of attaining the science orbit with a 360 m/s ΔV budget for apoapsis and periapsis adjustment. Monte Carlo analyses were also performed for a guidance system that modulates both bank angle and angle of attack with the reference set of uncertainties. An alpha and bank modulation guidance system reduces the 99.87 percentile ΔV 173 m/s (48%) to 187 m/s for the reference set of uncertainties.

NOMENCLATURE

AU   = Astronomical Unit
C_A = Aerodynamic axial force coefficient
C_N = Aerodynamic normal force coefficient
C.G.   = Center of Gravity
DOF   = degree of freedom
GRAM   = Global Reference Atmospheric Model
HYPAS   = Hybrid Predictor-corrector Aerocapture Scheme
JPL   = Jet Propulsion Laboratory
L/D   = Lift to drag ratio
LAURA   = Langley Aerodynamic Upwind Relaxation Algorithm
POST   = Program to Optimize Simulated Trajectories
SEP   = Solar electric propulsion
TPS   = Thermal protection system
ΔV = Velocity addition
σ = Standard deviation

BACKGROUND

NEPTUNE-TRITON EXPLORATION REFERENCE MISSION

The reference Neptune-Triton exploration mission was designed to provide Cassini and Galileo level exploration of the Neptune system.[2] The reference mission has a science orbiter to explore the Neptune-Triton system and two probes that enter Neptune's atmosphere 60 degrees apart in latitude. The science orbiter is placed in orbit about Neptune using an aerocapture maneuver described below. The science orbit can range between 3896 x 355000 km and 3896 x 500000 km. The aerocapture maneuver would be followed by propulsive maneuvers to place the spacecraft in a phasing orbit such that subsequent maneuvers would establish an orbit that would encounter Triton at regular intervals. Triton is then used as a tour engine to vary the orbit's inclination and line of apsides similar to Cassini's use of Titan. The reference orbiter would measure atmospheric, magnetic, and gravity characteristics and perform global imaging of both Triton and Neptune.

Launch dates studied for the Neptune-Triton Exploration mission range from 2016 to 2019. A February 21, 2017 launch date was chosen for this study with launch on a Delta IV 4050 Heavy launch vehicle inside a 5m fairing and a transit time of 10.25 years[3]. The launch spacecraft configuration consists of the orbiter, two entry probes, and a solar electric propulsion module. Five months prior to reaching Neptune, the two probes are released sequentially such that both probes' missions are completed before the orbiter reaches Neptune.[4] Four and a half months prior to reaching Neptune, a trajectory deflection maneuver is performed to target the entry interface point for aerocapture. Thirty minutes prior to entry interface the SEP is jettisoned. After atmospheric entry, the spacecraft executes the aerocapture maneuver described below to place the orbiter in the exploration orbit and begin a two or more year science mission.

1. Aerocapture Overview

Aerocapture is a form of aeroassist used to insert a spacecraft into a desired orbit at targets with an atmosphere. Aerocapture uses aerodynamic forces to dissipate the hyperbolic approach energy to an energy level needed to reach a target apoapsis after making a single pass through the atmosphere. An active guidance system must be used during the aeropass to compensate for uncertainties in entry flight path angles, atmospheric density profiles, and aerodynamics. After exiting the atmosphere, propulsive maneuvers are required to change the spacecraft's exit orbital elements to that of the desired phasing orbit. These maneuvers include a periapsis raise and any needed adjustments in apoapsis, inclination, and longitude of ascending node. The aerocapture maneuver is illustrated in Fig. 1. After the proper phasing with Triton is achieved, additional propulsive maneuvers are performed to provide the desired Triton encounter strategy. The ΔV required for the initial periapsis raise and assumed apoapsis for phasing was used as a performance metric in this study.

SIMULATION OF AEROCAPTURE ORBIT INSERTION

A high fidelity 3 DOF simulation of the aerocapture maneuver used to insert the spacecraft into its phasing orbit was developed in the program to optimize simulated trajectories, POST[9]. The aerocapture trajectory was simulated from the navigation delivery point, nominally 60 seconds prior to atmospheric interface, to atmospheric exit. The simulation determined the spacecraft's trajectory through Neptune's atmosphere and tracked key design parameters such as heat loads, deceleration loads, and ΔV required for the periapsis raise and apoapsis adjustments. The simulation was run in a Monte Carlo using uncertainties in the delivery point, spacecraft aerodynamics, and atmospheric density profiles to provide statistical data for the design parameters.

The simulation incorporated delivered states, aerodynamics, guidance, and control models specifically developed for the aeroshell and a model of Neptune's atmosphere as shown in Fig. 2. A brief description of each model follows.

ATMOSPHERE

An engineering type model of Neptune's atmosphere developed at Marshall Space Flight Center provided atmospheric state properties and composition.[6] The model, named Neptune-GRAM, is a global reference atmospheric model. Neptune-GRAM's state and composition properties were based on data from the Voyager flyby and stellar occultations.[6] Refer to reference 6 for a complete description of Neptune-GRAM.

Neptune's state property versus altitude relationships vary with latitude, season and time of day. These variations are represented in Neptune-GRAM by a parameter termed Fminmax. Fminmax ranges from –1 to +1 and is used to select a state property versus altitude profile for a particular latitude, season and time of day. Figure 3 shows the maximum range of density versus altitude profiles.

In the aerocapture simulation, the density versus altitude relationship was made to vary with latitude by making Fminmax a cosine function of latitude and was made to vary with season by adding a constant bias term, Fbias, to the latitudinal variation as shown in Eq. 1.

$$Fminmax = 0.44*\cos(4.0*latitude) + Fbias \qquad (1)$$

The range of Fbias was set to –0.56 to +0.56 so that Fminmax stayed within its –1 to +1 bounds. Figure 4

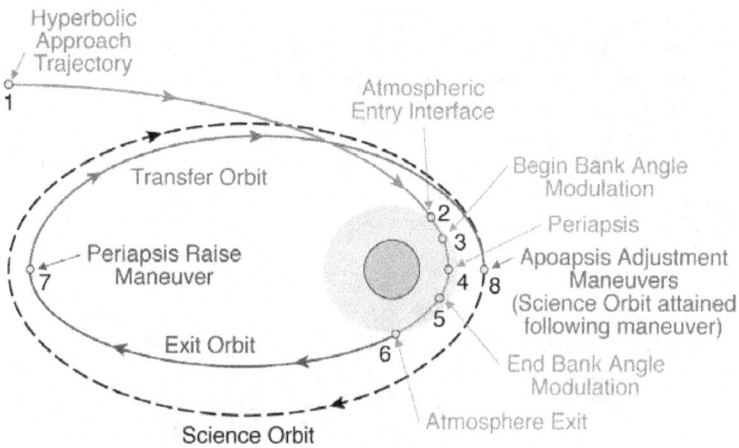

Figure 1. Illustration of aerocapture maneuvers.

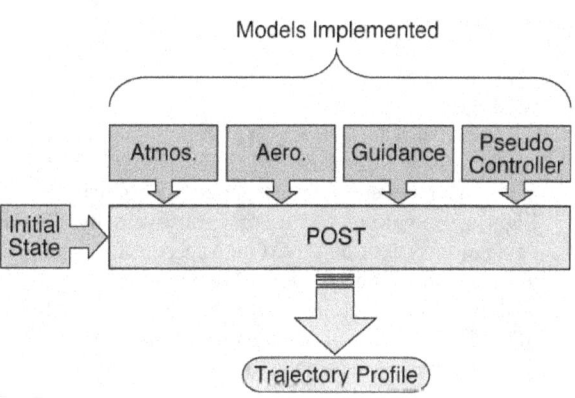

Figure 2. Models incorporated into POST simulation of Neptune aerocapture.

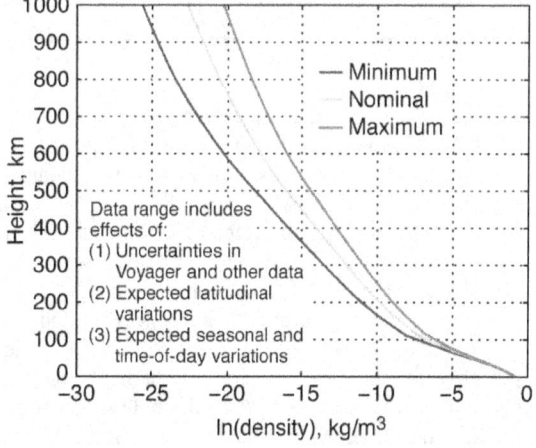

Figure 3. Neptune atmosphere model density profiles.

shows the variation of Fminmax with latitude.

Neptune-GRAM also superimposed high frequency perturbations onto the nominal atmospheric data to represent random variations in atmospheric properties. The magnitude of the perturbations can be scaled using the Neptune-GRAM parameter rpscale. A random perturbation seed value was used to generate randomly perturbed density values. Figure 5 shows randomly perturbed density versus altitude profiles for Fminmax = –1, 0 and +1.

AERODYNAMICS

A high fidelity aerodynamics model of the flat bottom ellipsled aeroshell was developed and incorporated into the simulation as an aerodynamic coefficient database. The database supplied axial and normal force coefficients as a function of angle of attack in the hypersonic flight regime.[7] The aerodynamic force coefficients were considered constant throughout the aeropass. Refer to reference 7 for more information regarding aerodynamics. The aeroshell geometry is shown in Fig. 6.

NAVIGATION

The Neptune-Triton Exploration Mission navigation model was provided by JPL. The navigation model determined vehicle entry states about a nominal –12.82° entry flight path angle and 29.0 km/s entry velocity. The modeled navigation system delivered the spacecraft to atmospheric interface with a 3σ dispersion of ±0.51° about the nominal entry flight path angle[5]. Refer to reference 5 for a complete description of the navigation model. Table 1 summarizes the navigation data used in the simulation.

Table 1.   Navigation Data

Entry Velocity, km/s	Entry Flight Path Angle, deg	Entry FPA Uncertainty, deg
29.0	-12.818	±0.5108

GUIDANCE

The Hybrid Predictor-corrector Aerocapture Scheme (HYPAS) aerocapture guidance algorithm developed at Johnson Space Center provided autonomous guidance for the simulation.[8] The HYPAS algorithm is an analytical control algorithm based on drag acceleration and altitude rate error for an aeropass through an exponential atmosphere.

In this study, two attitude control schemes were used by HYPAS. In the first, attitude control was limited to bank angle modulation. Bank angle controlled the rate of descent/ascent and effected drag through changes in atmospheric density. In this study, bank angle modulation was defined as the baseline guidance model. In the second attitude control scheme, attitude control included both angle of attack and bank angle modulation. Angle of attack modulation was

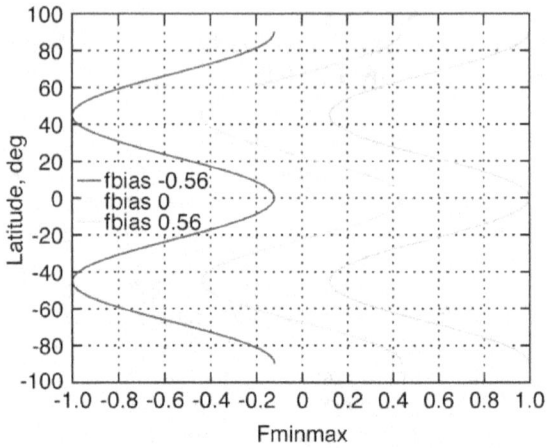

Figure 4. Latitudinal Variation of Fminmax.

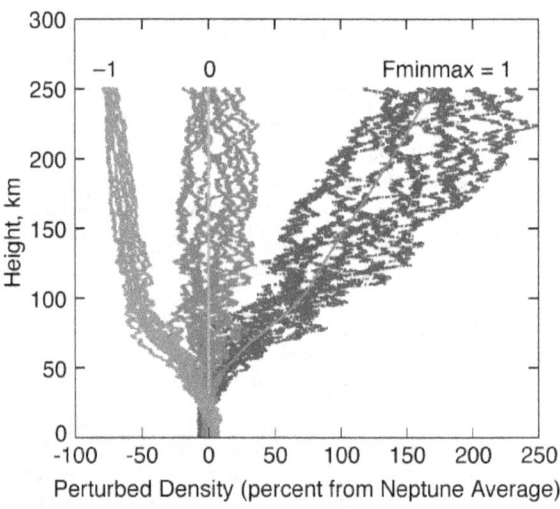

Figure 5. Sample Neptune-GRAM Monte-Carlo density output.

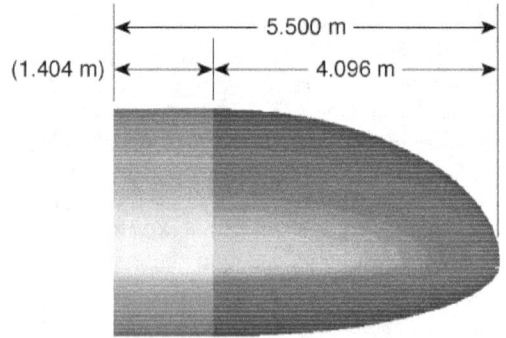

Figure 6. Aeroshell geometry.

used first to modulate the L/D and ballistic coefficient to meet the desired vertical L/D or exit velocity. If changes in ballistic coefficient were not sufficient to meet the targets then bank angle modulation was used to further effect drag. As the required L/D and drag changed due to high frequency density perturbations, angle of attack modulation was used to drive the bank angle to 90°. In this study, the combined bank and angle of attack modulation was defined as the alpha modulated guidance model. Both the baseline and advanced guidance models used roll reversals to maintain the wedge angle between the exit orbit and Triton's orbit to within 1°. Refer to reference 8 for complete descriptions of the bank modulated and alpha modulated guidance models.

CONTROL

A 3-DOF Pseudo controller developed at Langley Research Center was used to approximate the attitude dynamics of a 6 DOF system. The controller analytically calculated the time and angular travel required to reach the guidance commanded attitude. Once calculated, the controller ramped bank angle and/or angle of attack to the commanded value at a user specified maximum acceleration until the attitude rate reached a user specified maximum. The maximum acceleration and rates are defined such that the 3-DOF response is a good approximation of the 6-DOF system. This approach has provided good agreement with 6-DOF systems in previous simulations. Figure 7 shows the bank response to a bank command for the 3-DOF controller.

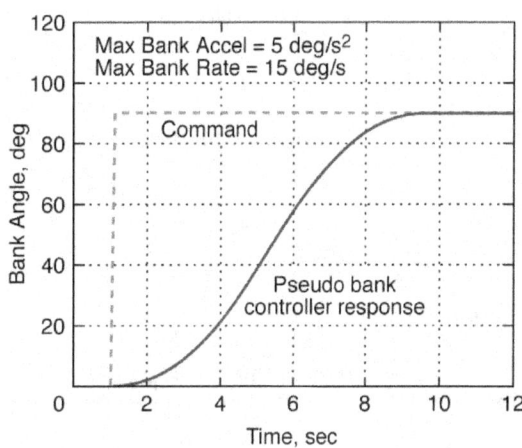

Figure 7. Bank Response of 3-DOF Bank Controller

MONTE CARLO ANALYSIS

The vehicle performance was quantified by statistical data from Monte Carlo Analyses. The analyses consisted of 2000 individual Neptune aerocapture simulations with random perturbations in arrival states, vehicle aerodynamics and Neptune's atmosphere. A Monte Carlo executive script created simulation input files with generated perturbations and coordinated simultaneous execution of the simulations on multiple processors across multiple computers. Various post processing scripts were used to determine the statistical parameters for the 2000 simulations in each Monte Carlo analysis and to generate plots.

The Monte Carlo analyses consisted of a reference case and three sensitivity case studies. In the reference case, delivery, atmospheric and aerodynamic uncertainties were based on state of the art navigation, current knowledge of Neptune atmosphere and computational fluid dynamics analyses respectively. Table 2 lists the uncertainties and distribution types used in the Monte Carlo reference case. In the first sensitivity case study, the magnitudes of the high frequency random density perturbations were reduced by 50%. In the second sensitivity study, the latitudinal variation of Fminmax was removed and the uncertainty in mean density increased. In the third sensitivity study, alpha modulation was added to the reference bank modulation approach. The Monte Carlo analyses performed are summarized in Table 3.

Table 2. Monte Carlo Uncertainties

Category	Variable	Nominal	±3σ or min/max	Distribution
Delivery State				
	X position	19813.3 km	From covariance	Correlated
	Y position	-16908.2 km	From covariance	Correlated
	Z position	2612.7 km	From covariance	Correlated
	X velocity	-22.953 km/s	From covariance	Correlated
	Y velocity	-13.324 km/s	From covariance	Correlated
	Z velocity	11.316 km/s	From covariance	Correlated
Atmosphere				
	Random Pertubation seed	1	1 to 9999	Uniform
	Fbias	0	-0.56 to 0.56	Uniform
Aerodynamics				
	Trim angle of attack	40.0	±4.0	Normal
	C_A	0.349	±0.048	Uniform
	C_N	1.771	±0.120	Uniform
Mass Properties				
	Axial C.G. (Xcg/L)	0.51	±0.50%	Uniform
	Radial C.G. (Zcg/L)	-0.0166	±0.125%	Uniform

Table 3. Summary of Monte Carlo Analyses

Case	Perturbation Scale	Fminmax	Guidance
Reference	1.0	f(latitude)	Bank angle modulation
Reduced Density Perturbations	0.5	f(latitude)	Bank angle modulation
Increased Density Uncertainty	1.0	global	Bank angle modulation
Alpha Modulation	1.0	f(latitude)	Bank + alpha modulation

RESULTS

REFERENCE CASE

In the reference case, the reference set of uncertainties were used in the Monte Carlo analyses. Figure 8 shows the reference case atmospheric density variation in the aerocapture altitudes as a ratio of perturbed density to nominal density. The reference case density varies up to a factor of 2.25.

The guidance compensates for these variations as well as delivery, aerodynamic, and C.G. variations by using the spacecraft's available control authority. Dispersions in apoapsis altitude at atmospheric exit result when the spacecraft's control authority is insufficient to compensate for the variations. Figure 9 shows the reference case dispersion in apoapsis and periapsis altitude and Fig. 10 shows a histogram of exit apoapsis altitude.

For the reference uncertainties, 100% of the cases successfully captured and 66.1% of the cases were within the desired science orbit apoapsis bounds. The dispersion in apoapsis altitude between the 0.13 percentile and 99.87 percentile was 461.4E+03 km. The apoapsis percentiles are summarized in Table 4.

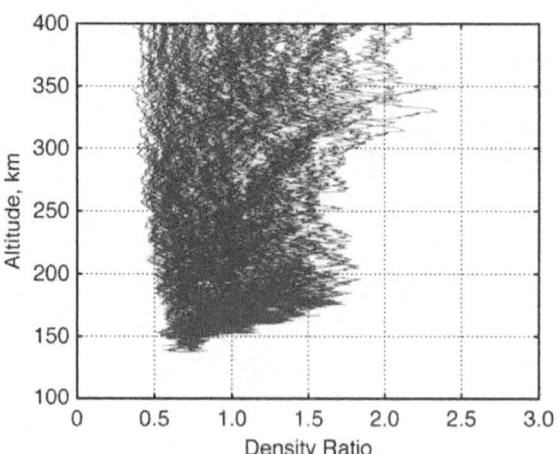

Figure 8. Reference Case Atmospheric Density Variation

Table 4. Apoapsis Percentiles – Reference Case

Apoapsis Statistics	Altitude, x10³ km
0.13 percentile	371.3
50.00 percentile	477.9
99.87 percentile	832.7

Impulsive maneuvers were used to adjust the atmospheric exit orbit to a nominal 3986 x 430000 km orbit. The 99.87 percentile ΔV required to attain the nominal orbit and the percentage of cases captured were used as performance and robustness metrics. The 99.87 percentile ΔV was used rather than a 3σ value since the ΔV distribution was skewed. The skewing is a result of ΔV being required for periapsis raise in all cases regardless of whether or not the target apoapsis is met. Any error in apoapsis only results in increased ΔV. Table 5 summarizes these metrics for the reference case.

Table 5. Performance Metrics – Reference Case

Robustness Statistics	%
Cases Captured	100
Cases within Target Bounds	66.1
Performance Statistics	**ΔV, m/s**
0.13 percentile	88
50.00 percentile	141
99.87 percentile	360

The system is sufficiently robust to overcome reference delivery, atmospheric and aerodynamic uncertainties. A 99.87 probability of attaining the science orbit is possible with a 360 m/s ΔV budget for periapsis raise and apoapsis adjustment. The reference case ΔV histogram is shown in Fig. 11.

REDUCED ATMOSPHERIC HIGH FREQUENCY DENSITY PERTURBATIONS

In the reduced high frequency density perturbation sensitivity study, the magnitude of atmospheric high frequency perturbations were reduced 50% using the GRAM perturbation multiplier rpscale. The reduction may be possible with improved knowledge of Neptune's atmosphere. A half scale density perturbation near the periapsis of a selected aeropass is shown relative to full scale in Fig. 12. Note that a random perturbation is equally likely to be below the mean density as above it. Figure 13 shows the dispersion in apoapsis and periapsis altitude and Fig. 14 shows a histogram of exit apoapsis altitude.

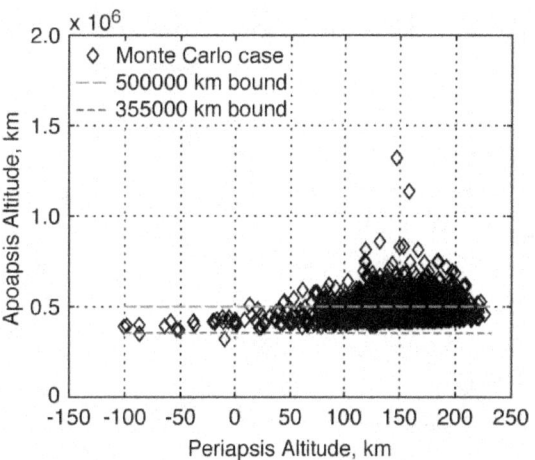

Figure 9. Apoapsis and Periapsis Altitude Dispersion – Reference Case

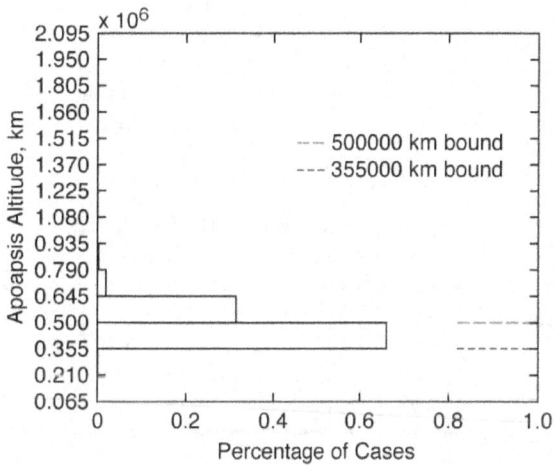

Figure 10. Apoapsis Altitude Histogram – Reference Case

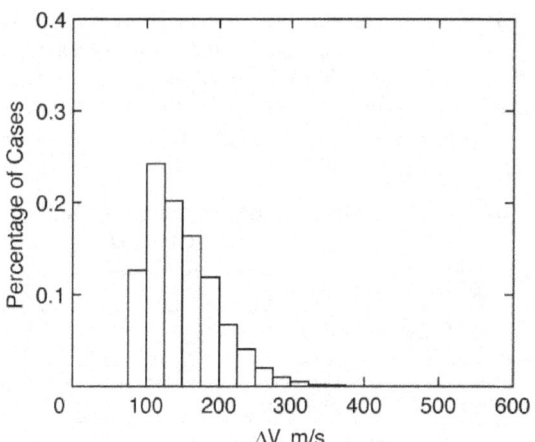

Figure 11. ΔV Histogram – Reference Case

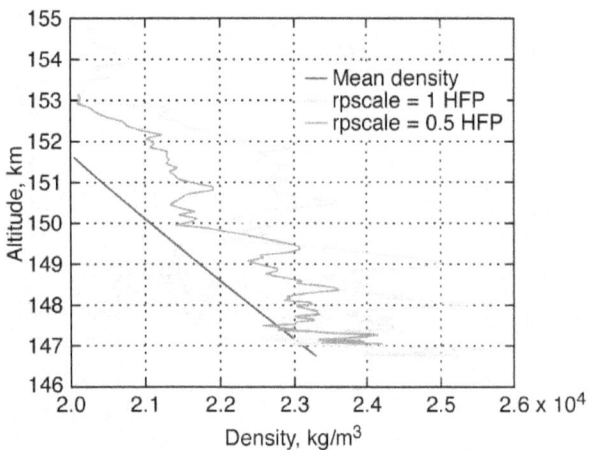

Figure 12. Comparison of Full and Half Scale High Frequency Density Perturbations

For an atmosphere with half scale high frequency density perturbations, 100% of the cases successfully captured and 81.8% of the cases were within the apoapsis bounds. In addition, the dispersion in apoapsis altitude was reduced 221.2E+03 km (47.9%) to 240.2E+03 km. The apoapsis percentiles are summarized in Table 6.

Table 6. Apoapsis Percentiles – Half Scale High Frequency Density Perturbations

Apoapsis Statistics	Altitude, x10³ km
0.13 percentile	412.7
50.00 percentile	456.3
99.87 percentile	652.9

The performance statistics are summarized in Table 7. Reducing high frequency density perturbations 50% reduces 99.87 percentile ΔV to 271 m/s, an 89 m/s (24.7%) reduction relative to the reference case. The smaller density perturbations can be compensated for with less control authority. This allows the guidance to improve targeting of the science orbit apoapsis and reduce ΔV needed for apoapsis adjustment. The ΔV histogram is shown in Fig. 15.

Table 7. Performance Metrics – Half Scale High Frequency Density Perturbations

Robustness Statistics	%
Cases Captured	100
Cases within Target Bounds	81.8
Performance Statistics	**ΔV, m/s**
0.13 percentile	87
50.00 percentile	118
99.87 percentile	271

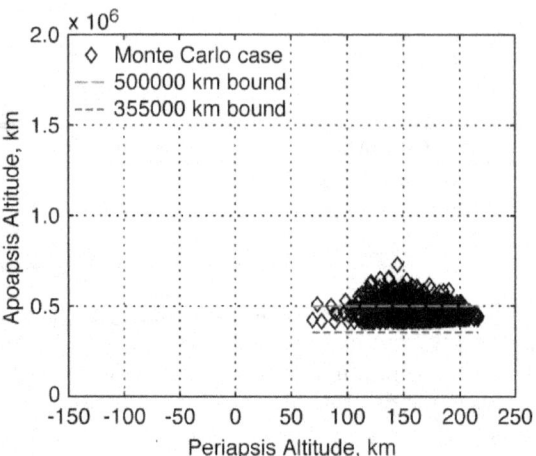

Figure 13. Apoapsis and Periapsis Altitude Dispersion – Half Scale High Frequency Density Perturbations

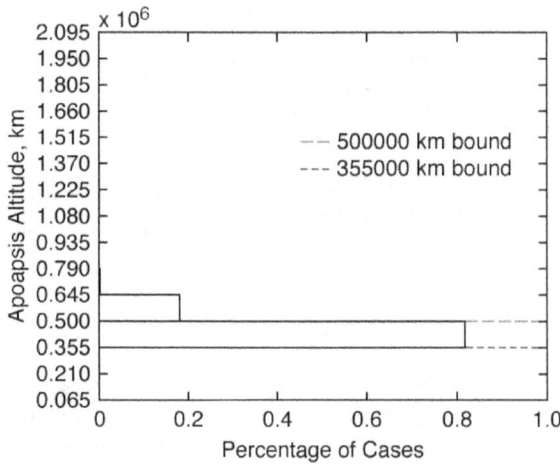

Figure 14. Apoapsis Altitude Histogram – Half Scale High Frequency Density Perturbations

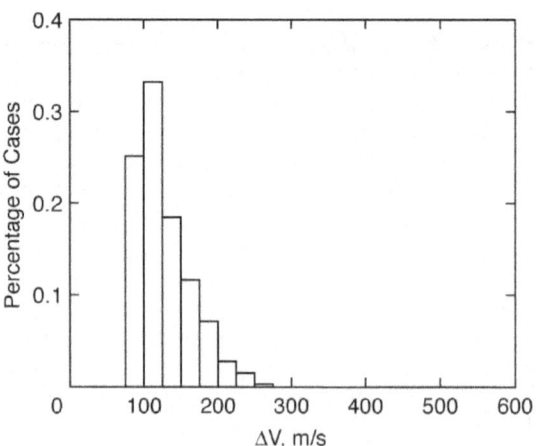

Figure 15. ΔV Histogram – Half Scale High FrquencyDensity Perturbations

INCREASED MEAN DENSITY UNCERTAINTY

In this sensitivity study, the latitudinal variation of Fminmax was removed. In addition, the Monte Carlo range of Fbias was increased from $-0.56 \leq$ Fbias ≤ 0.56 to $-1 \leq$ Fbias $\leq +1$. This made Fminmax constant throughout a given aeropass and resulted in the largest possible uncertainty range of Fminmax, -1 to $+1$. It also produces the maximum uncertainty in mean density from one aeropass to another. The maximum range of Fminmax is 33% larger than the Fminmax range of the reference case. Figure 16 shows the density ratio in the aerocapture altitudes. In the altitude range of 125 km to 300 km where the majority of hyperbolic approach velocity is dissipated, the density variations above the nominal are approximately 25% larger than the reference atmosphere while density variations below the nominal are approximately 50% larger reductions.

The larger reduction in atmospheric density impacts the guidance's ability to target the apoapsis. For cases in which the density decreases, the spacecraft does not have the control authority to overcome the centripedal acceleration and descend into more dense atmospheric regions. As a result it exits with an apoapsis above the bounds. The dispersion in apoapsis and periapsis altitude is shown in Fig. 17. Figure 18 shows a histogram of exit apoapsis altitude.

For an atmosphere with increased uncertainties in mean density, 100% of the cases successfully captured and 76.3% of the cases were within the apoapsis bounds. The dispersion in apoapsis altitude increased 498.7E+03 km (108.1%) to 960.0E+03 km. The apoapsis percentiles are summarized in Table 8.

Table 8. Apoapsis Altitude Percentiles – Increased Mean Density Uncertainty

Apoapsis Statistics	Altitude, x10^3 km
0.13 percentile	324.9
50.00 percentile	448.7
99.87 percentile	1284.5

The system is robust enough to capture 100% of the cases with increased density uncertainty. However, the 99.87 percentile ΔV increased 96.0 m/s to 456 m/s, a 26.7% increase relative to the reference case. Table 9 summarizes the performance statistics. The ΔV distribution is shown in Fig. 19.

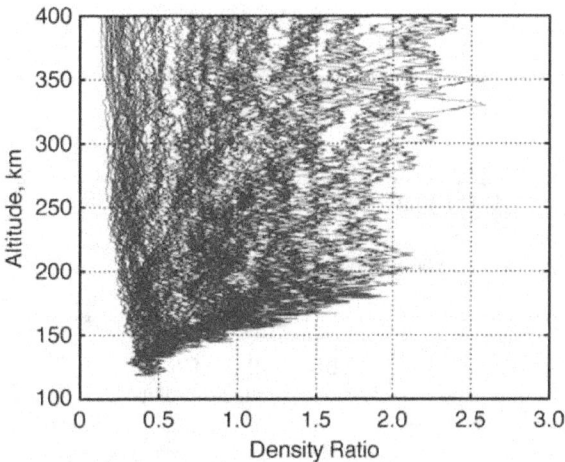

Figure 16. Atmospheric Density Variation for Maximum Range of Fminmax

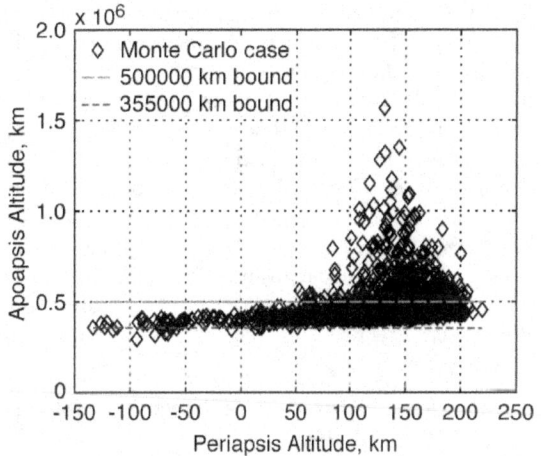

Figure 17. Apoapsis and Periapsis Altitude Dispersion – Increased Mean Density Uncertainty

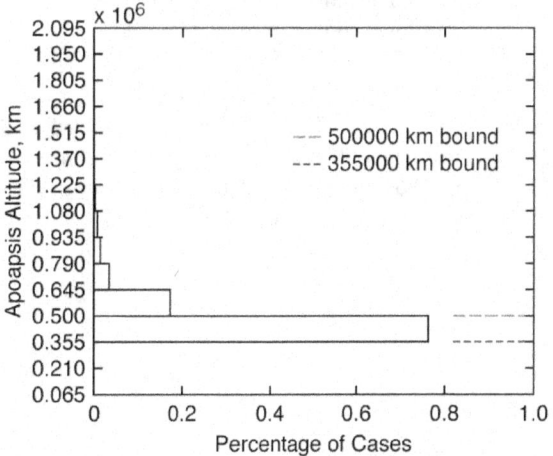

Figure 18. Apoapsis Altitude Histogram – Increased Mean Density Uncertainty

Table 9. Performance Metrics – Increased Mean Density Uncertainty

Robustness Statistics	%
Cases Captured	100
Cases within Target Bounds	76.3
Performance Statistics	**Δ V, m/s**
0.13 percentile	88
50.00 percentile	125
99.87 percentile	456

ALPHA MODULATED GUIDANCE

In the alpha modulated guidance sensitivity study, a guidance algorithm with angle of attack modulation in addition to bank modulation was used. The reference atmospheric uncertainty assumptions were used with the alpha modulated guidance case. The alpha modulated guidance improves the spacecraft's ability to compensate for density perturbations through changes in drag. Angle of attack modulation changes the drag more quickly than banking thus allowing the spacecraft to respond more quickly to density perturbations. The dispersion in apoapsis and periapsis altitude is shown in Fig. 20. Figure 21 shows a histogram of exit apoapsis altitude.

For a spacecraft with the alpha modulated guidance system, 100% of the cases were successfully captured and 98.4% of the cases were placed within the apoapsis bounds. The dispersion in apoapsis altitude decreased 335.5E+03 km (72.7%) to 125.9E+03 km relative to the reference case. The apoapsis percentiles are summarized in Table 10.

Table 10. Apoapsis Altitude Percentiles – Alpha Modulated Guidance

Apoapsis Statistics	Altitude, x10^3 km
0.13 percentile	402.6
50.00 percentile	425.1
99.87 percentile	528.5

The alpha modulated guidance performance statistics are summarized in Table 11. The alpha modulated guidance reduced the 99.87 percentile ΔV by 174.0 m/s (48.3%) to 186 m/s relative to the reference case. The ΔV histogram is shown in Fig. 22.

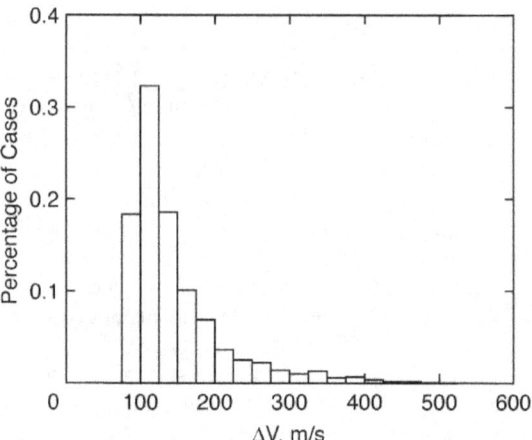

Figure 19. ΔV Histogram – Increased Mean Density Uncertainty

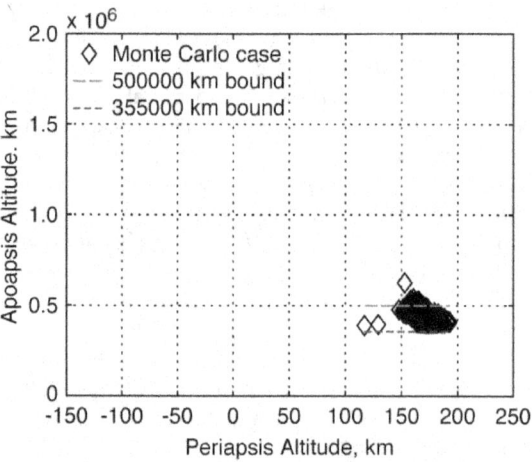

Figure 20. Apoapsis and Periapsis Altitude Dispersion – Alpha Modulated Guidance

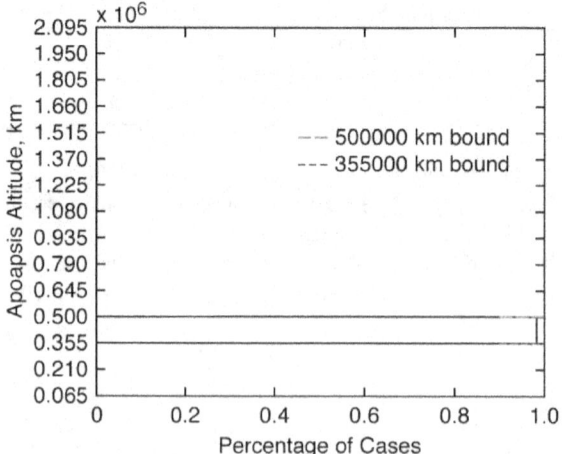

Figure 21. Apoapsis Altitude Histogram – Alpha Modulated Guidance

Table 11. Performance Metrics – Alpha Modulated Guidance

Robustness Statistics	%
Cases Captured	100.0
Cases within Target Bounds	98.4

Performance Statistics	Δ V, m/s
0.13 percentile	88
50.00 percentile	117
99.87 percentile	186

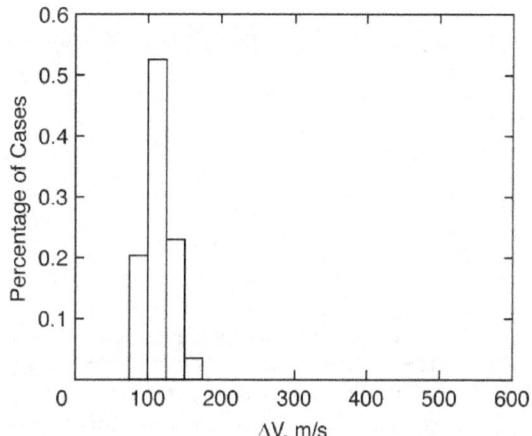

Figure 22. ΔV Histogram – Alpha Modulated Guidance

CONCLUSION

The performance analysis has shown that for the current Neptune atmospheric model, the 0.806 L/D flatbottom ellipsled spacecraft is a viable design that captures 100% of the cases and has a 99.87 probability of successfully inserting the orbiter into its science orbit with a 360 m/s ΔV budget for periapsis raise and apoapsis adjustment.

The analysis has also shown that the design is viable for atmospheric models with different variability assumptions. For an atmosphere with Fminmax uncertainties 33% larger than that of the reference atmospheric model, the spacecraft can attain the science orbit with a 99.87 probability given a 456 m/s ΔV budget for periapsis and apoapsis adjustment. If a better understanding of Neptune's atmosphere leads to a 50% reduction in high frequency density perturbation magnitude, the science obit can be attained with a 99.87 probability given a 271 m/s ΔV budget for periapsis and apoapsis adjustment. The analysis has also shown that with a combined angle of attack and bank angle modulation the spacecraft can attain the science orbit with a 99.87 probability given a 186 m/s ΔV budget for periapsis and apoapsis adjustment.

REFERENCES

[1]Lockwood, M. K., "Neptune Aerocapture Systems Analysis," AIAA-2004-4951, Conference proceedings of the AIAA Atmospheric Flight Mechanics Conference and Exhibit, Providence, Rhode Island, August 2004.

[2]Spilker, Tom, "Significant Scienceat Titan and Neptune," EGU04-A-2270, Conference Proceedings of European Geosciences Union Planetary and Solar Sciences Programme, Nice, France, April 2004

[3]Noca, Muriel, Bailey, R. W., "Mission Trades for Aerocapture at Neptune," AIAA-2004-3843, Conference Proceedings of the 40th AIAA/ASME/SAE/ASEE Joint Propulsion Conference and Exhibit, Huntsville, AL, July 2004

[4]Bailey, R. W., Hall, J. L., Spilker, T. R., Okong'o, N. O., "Neptune Aerocapture Mission and Spacecraft Design Overview" AIAA-2004-3842, Conference Proceedings of the 40th AIAA/ASME/SAE/ASEE Joint Propulsion Conference and Exhibit, Huntsville, AL, July 2004

[5]Hanak, C., Bishop R., "Aerocapture Navigation Analysis at Titan and Neptune," AIAA-2004-5180, Conference Proceedings of the AIAA Atmospheric Flight Mechanics Conference and Exhibit, Providence, Rhode Island, August 2004

[6]Justus, C. G., Duvall, A., Keller, V., "Atmospheric Models for Aerocapture Systems Studies," AIAA-2004-4952, Conference Proceedings of the AIAA Atmospheric Flight Mechanics Conference and Exhibit, Providence, Rhode Island, August 2004

[7]Edquist, K. T., Hoffman, D. A., Rea, J.R., "Configuration, Aerodynamics and Stability Analysis for a Neptune Aerocapture Orbiter," AIAA-2004-4953, Conference Proceedings of the AIAA Atmospheric Flight Mechanics Conference and Exhibit, Providence, Rhode Island, August 2004

[8]Masciarelli, J. P., Westhelle, C. H., Graves, C. A., "Aerocapture Guidance Performance for the Neptune Orbiter," AIAA-2004-4954, Conference Proceedings of the AIAA Atmospheric Flight Mechanics Conference and Exhibit, Providence, Rhode Island, August 2004

[9]Bauer, G. L., Cornick, D. E., Olson, D. W., Petersen, F. M., Stevenson, R., "Program to Optimize Simulated Trajectories (POST II)," NASA CR-2770, February 1977

AEROCAPTURE GUIDANCE PERFORMANCE FOR THE NEPTUNE ORBITER

James P. Masciarelli and Carlos H. Westhelle,
NASA Johnson Space Center, Houston, Texas, 77058

Claude A. Graves
NASA Johnson Space Center, Houston, Texas, 77058

A performance evaluation of the Hybrid Predictor-corrector Aerocapture Scheme (HYPAS) guidance algorithm for aerocapture at Neptune is presented in this paper for a Mission to Neptune and the Neptune moon Triton[1]. This mission has several challenges not experienced in previous aerocapture guidance assessments. These challengers are a very high Neptune arrival speed, atmospheric exit into a high energy orbit about Neptune, and a very high ballistic coefficient that results in a low altitude acceleration capability when combined with the aeroshell L/D. The evaluation includes a definition of the entry corridor, a comparison to the theoretical optimum performance, and guidance responses to variations in atmospheric density, aerodynamic coefficients and flight path angle for various vehicle configurations (ballistic numbers). The benefits of utilizing angle-of-attack modulation in addition to bank angle modulation to improve flight performance is also discussed. The results show that despite large sensitivities in apoapsis targeting, the algorithm performs within the allocated ΔV budget for the Neptune mission using only bank angle modulation. The addition of angle-of-attack modulation with as little as ±5 degrees of amplitude significantly improves the accuracy in final orbit apoapsis. Although angle-of-attack modulation complicates the vehicle design its performance enhancement reduces aerocapture risk and reduces the propellant consumption needed to reach the high energy target orbit.

Nomenclature

c	=	Convective heat transfer coefficient
C_D	=	Aerodynamic drag coefficient
D	=	Drag acceleration
L	=	Lift acceleration
m	=	Vehicle mass
M	=	Planetary body mass
$\dot{q}_c$	=	Convective aerodynamic heating rate
r	=	Radial distance from center of planetary body
R_N	=	Aeroshell nose radius
A	=	Aerodynamic reference area
V	=	Relative velocity
ΔV	=	Change in velocity
γ	=	Inertial flight path angle
V_i	=	Inertial velocity
μ	=	Gravitational constant
ρ	=	Atmosphere density
σ	=	Bank angle

INTRODUCTION

The objective of the Neptune mission is to explore Neptune and one of its moons, Triton, in similar fashion to the Cassini-Huygens Mission. In addition to performing atmospheric observations and detailed gravity and magnetic field determination via two atmospheric entry probes and one orbiting vehicle, the final orbit of the vehicle is placed in a nominal 3,896 x 430,000 km orbit about the planet in order to allow observations of Triton, one of the moons of Neptune. Aerocapture is an efficient way to provide the energy dissipation from a high energy approach trajectory needed to enter into a low energy orbit about a planet during one pass through a planet's atmosphere. After the pass through the atmosphere, only a relatively small propulsive maneuver is required to place the spacecraft into the desired final orbit.

Aerocapture is considered enabling for this Neptune mission and provides significant reduction in mass or trip time for other planetary destinations[2,3]. This technique allows a spacecraft to capture into a desired orbit about a given body at significantly lower ΔV cost and thus provides an overall mass and cost savings to the mission. The Hybrid Predictor-corrector Aerocapture Scheme (HYPAS)[4] is a candidate for the aerocapture guidance algorithm. This algorithm has been analyzed extensively for aerocapture missions at Earth, Mars, and Titan, however aerocapture for this Neptune mission has challenges not encountered in previous aerocapture assessments. This paper addresses the challenges associated with the aerocapture mission at Neptune and defines the performance of the algorithm, including its sensitivity to perturbations in aerodynamics, atmospheric density, and the arrival state vector.

In the reference mission, the vehicle enters the atmosphere at an altitude of 1,000 km with an inertial velocity of 29.0 km/s. The vehicle uses a blunt body flat bottom ellipsled aeroshell[5], see Figure 1, with a ballistic number (m/C_DA) of 895 kg/m^2, and a lift-to-drag ratio (L/D) of 0.8. The aerocapture guidance algorithm targets the vehicle to a desirable atmospheric exit state vector that allows it to reach the targeted 430,000 km apoapsis.

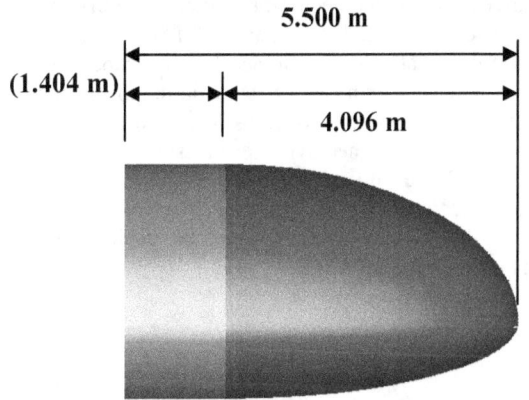

Figure 1. Aeroshell Geometry

NEPTUNE AEROCAPTURE CHALLENGES

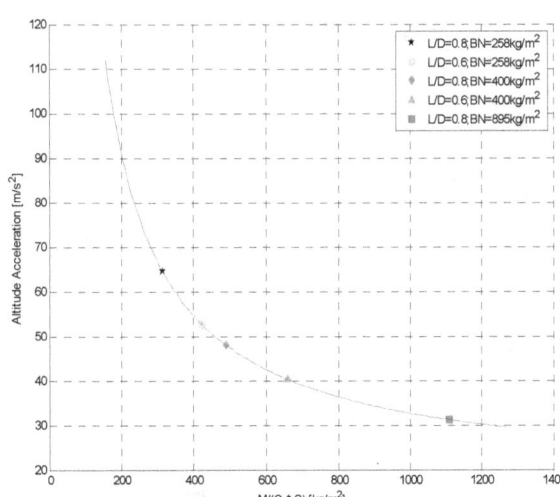

Figure 2. Altitude Acceleration capability vs. M/(C$_L$A)

The nature of the Neptune mission presents a set of new challenges to the guidance system. The low-thrust Solar Electric Power (SEP) trajectory brings the vehicle after a 10.25 year transit time to an entry velocity of 29.0 km/s; compared to typical missions to inner planets that present an entry velocity of approximately 11.0 km/s or lower, and earth return from planetary missions that may reach 13 km/s for a return from Mars. This high entry velocity associated with the Neptune mission causes the vehicle to experience large aerodynamic and thermodynamic loads that constrain the entry corridor.

The second challenge results from the high value of M/C_LA for the given vehicle design. Previous aerocapture missions used vehicles with M/C_LA of 400 kg/m^2 to 700 kg/m^2. Mass and packaging considerations for the Neptune mission result in a M/C_LA of more than 1,100 kg/m^2. M/C_LA is an inverse measure of control authority for altitude acceleration that is essential for efficient control of the drag acceleration profile. As can be seen in Figure 2, the altitude acceleration capability decreases rapidly as the M/C_LA of the vehicle increases. The curve has been generated using the equation for normal to the velocity vector acceleration

$$\ddot{h} = \bar{q} \cdot \frac{C_L S}{m} \cdot \cos \sigma - \left[\frac{\mu}{R^2} - \frac{V_i^2}{R} \right] \cos \gamma, \tag{1}$$

where, $\mu = 6.8713e6$ km^3/s^2, R = 24981705.48 m, V = 26872.0824 m/s and $\bar{q}_{max}$ = 14.753 kPa. As altitude acceleration capability decreases, it becomes more difficult for the vehicle to respond to random density perturbations in a timely manner.

The data was taken from the maximum dynamic pressure point in the nominal trajectory for a L/D of 0.8 and ballistic number of 258 kg/m^2 configuration. For simplification, the flight path angle, γ, and bank angle, σ, are assumed to be 0 deg. It can be seen that for a given point in a trajectory, the final configuration (L/D of 0.8 and ballistic number of 895 kg/m^2) provides only half the altitude acceleration capability as the counterpart with a lower ballistic number. This lack of control authority, will thus force the guidance system to seek a higher dynamic pressure, and thus force the vehicle into a steeper trajectory. While the steeper trajectory aids the guidance control, it also increases the thermodynamic loads on the vehicle, and is thus undesirable.

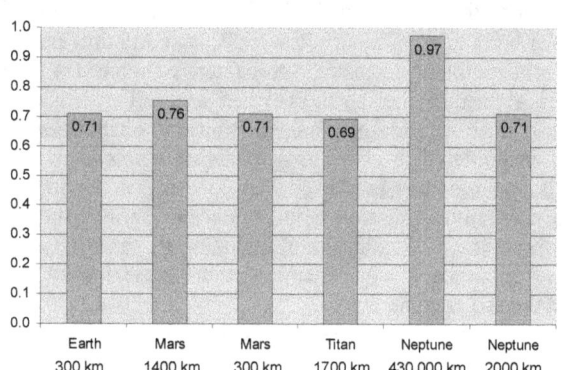

Figure 3. Ratio of Exit Velocity to Escape Velocity

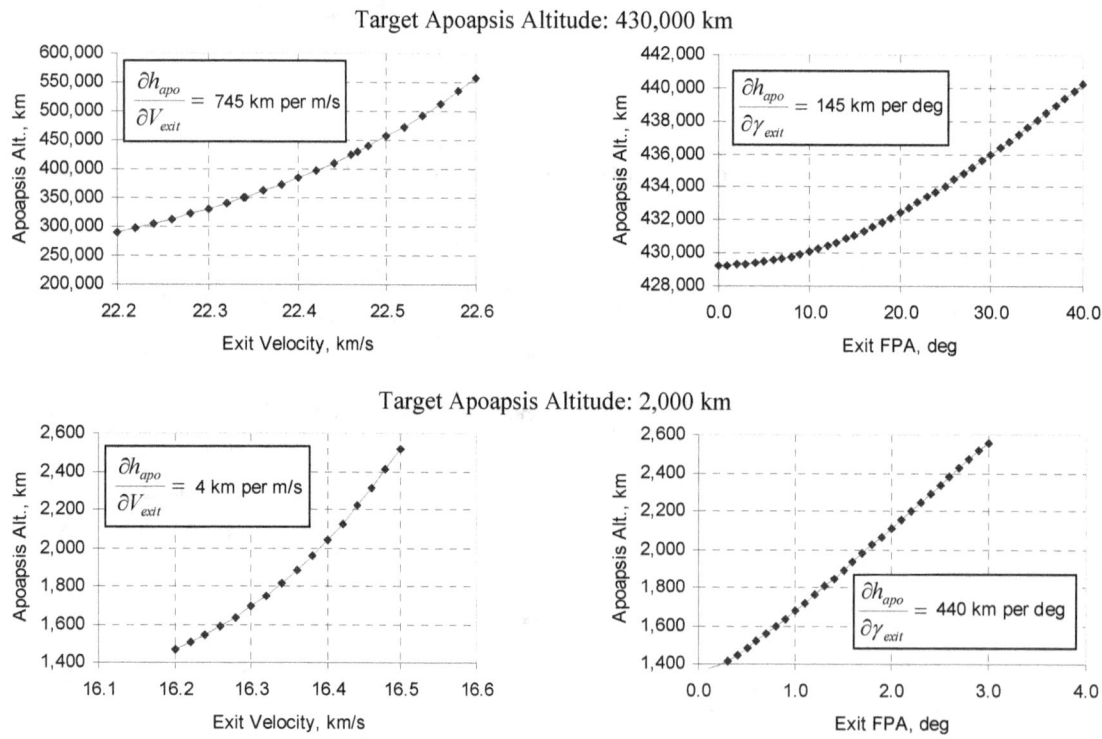

Figures 4-7. Sensitivity to Atmospheric Exit Conditions

100

The third new challenge results from the high energy final target orbit of 3,896 x 430,000 km that requires an exit velocity near 22.5 km/s or about 97% of the Neptune escape speed. Figure 3 shows typical exit-to-escape velocity ratios for missions using aerocapture at other destinations with atmospheric exit at near orbital speed or about 71% of escape speed. The very high exit speed for this Neptune mission results in a large sensitivity of the post aerocapture orbit to the atmospheric exit conditions. This is illustrated in Figures 4-7, which show the effects of exit velocity error and exit flight path angle on apoapsis altitude as a function of the target apoapsis altitude for two different ranges of apoapsis altitudes that correspond to low and high energy planetary orbits. While the sensitivity to exit flight path angle is slightly larger for the 2,000 km target apoapsis altitude, a 1 m/s error in exit velocity yields a 745 km error in apoapsis altitude for the 430,000 km target versus a 4 km error for the 2000 km target, about two orders of magnitude larger.

DESCRIPTION OF TRAJECTORY SIMULATION

The Simulation and Optimization of Rocket Trajectories (SORT)[6] program was used to perform the numerical trajectory simulation. SORT is a multi-purpose computer program that uses three degrees-of-freedom to simulate the translation flight dynamics of an arbitrary aerospace vehicle about a given central body with phase plane logic to simulate the bank dynamics. The vehicle is assumed to be in an aerodynamically trimmed attitude. The simulation interface allows the user to easily adjust vehicle parameters, planet geometry, gravity model, and atmospheric model.

For the Neptune aerocapture trajectory simulation, the planet and gravity model used are based on a spherical body of radius 24,764 km with an inverse square gravity field with a gravitational parameter of 6.8713×10^6 km^3/s^2. The atmosphere model used for the trajectory simulation is the Neptune Global Reference Atmosphere Model (GRAM)[7]. Atmospheric density is computed as a function of altitude, and the engineering model allows a variation of the density profile between a minimum and maximum expected value that is includes the effects of latitude, seasonal, and diurnal effects. The parameter in the model that allows for this control (FMINMAX) can be varied continuously between -1.0 and +1.0, corresponding to the minimum and maximum expected mean density, respectively. A value of 0.0 produces a nominal density profile. The Neptune GRAM also has the capability to superimpose random density perturbations on the selected profile to model the relatively high frequency density gradients resulting from gravity waves. The simulation tool uses a Sutton-Graves[8] stagnation point convective heat rate equation,

$$ \dot{q}_c = cV^3 \sqrt{\frac{\rho}{R_N}}, \tag{2} $$

where $\dot{q}_c$ is the heating rate, V is the atmospheric relative velocity, ρ is the atmospheric density, R_N is the vehicle's nose radius, and c is the Sutton-Graves constant, which for Neptune is 6.96×10^{-9} kg$^{0.5}$/m.

The atmospheric entry conditions were taken from the proposed Neptune Triton Exploration Mission[1]. This mission results in an inertial entry velocity of 29 km/s at 1000 km altitude.

The Neptune entry vehicle uses bank angle to control the in-plane dynamics and bank reversal to control the trajectory inclination. This Neptune aerocapture assessment used the Hybrid Predictor-corrector Aerocapture Scheme (HYPAS) algorithm to develop the bank angle magnitude and bank reversal commands. This guidance algorithm was originally developed at the Johnson Space Center in mid 1980's as the guidance algorithm for the Aeroassist Flight Experiment (AFE). This analytically derived guidance algorithm uses drag based deceleration and altitude rate errors as feedback terms in the closed loop guidance operation. The algorithm has been extensively used in various different simulation studies of aerocapture into low energy planetary orbits, with a range of different initial state vectors, vehicle ballistic numbers, and lift-to-drag ratios, as well as multiple destinations, such as Earth, Mars, and Titan[9]. Guidance initialization constants and the feedback gains are used to control the profile shape and responsiveness to perturbations to achieve the desired dynamic pressure, heat rate, deceleration, and the total theoretical corridor capture.

The HYPAS guidance algorithm is divided into two flight phases, an equilibrium glide, and an exit phase. In the first phase, the algorithm generates bank angle commands to drive the vehicle towards a balance of lift, gravity, and centripetal forces, or equilibrium glide conditions to assure capture and control levels of loads. Once the vehicle has decelerated to a specified velocity, the exit phase is initiated. In this phase, the guidance computes a predicted velocity vector at atmospheric exit altitude, based on analytically derived equations. The bank angle is commanded to adjust the exit state so that the vehicle can achieve the desired target apoapsis. The robustness of the algorithm stems from the ability to individually tune the initialization parameters for both phases, allowing for a controlled and stable capture, and maximum performance during exit.

PERFORMANCE FOR NEPTUNE AEROCAPTURE

The first step in the performance analysis of the guidance system involves determination of the theoretical entry flight path angle corridor. This corridor is used to define the limits on the arrival flight path angle as well as to define the nominal, or target, entry flight path angle. An optimum aerocapture flight profile is developed for this entry corridor. Development of this profile is a multi-step process of defining the aerocapture guidance initialization parameters and testing the effect on variations to entry flight path angle, aerodynamic coefficients, and atmospheric density throughout the entry corridor. The final step in the analysis is a Monte Carlo simulation; to assess the overall performance of the guidance algorithm under expected random flight conditions.

THEORETICAL ENTRY CORRIDOR

Defining the theoretical entry corridor allows for a definition of the desired nominal entry flight path angle as well as the limitations on navigation errors of delivery. The steep side of the corridor is defined as the steepest entry flight path angle which allows the vehicle to just reach the target apoapsis while flying lift vector up over the entire duration of the flight. The shallow side of the corridor is defined as the shallowest entry flight path angle which allows the vehicle to just reach the target apoapsis while flying lift vector down over the entire duration of the flight.

For this assessment, the theoretical entry flight path angle corridor was determined by using the Neptune GRAM with the nominal, minimum, and maximum atmospheric density profiles to account for the uncertainty in the density profile (see Table 1). A combined corridor was developed that includes the effects of atmospheric density variations and this results in an entry corridor of 1.78 deg. This corridor is sufficiently large enough to accommodate the expected delivery errors. The nominal entry flight path angle was chosen to be –12.77 deg, the middle of this combined theoretical corridor.

Atmosphere	Steep Side (deg)	Shallow Side (deg)	Width (deg)	Middle (deg)
Nominal	-13.81	-11.49	2.32	-12.65
Minimum	-13.96	-11.88	2.09	-12.92
Maximum	-13.66	-11.08	2.58	-12.37
Combined	-13.66	-11.88	1.78	-12.77

Table 1. Theoretical Entry Flight Path Angle

AEROCAPTURE FLIGHT PROFILE

Once the theoretical entry corridor is defined, an optimum performance for the guidance algorithm can be determined. By having the vehicle fly full lift up as it enters the atmosphere and then command the vehicle to bank to a full lift down attitude at the correct instant in order to achieve the precise target apoapsis altitude, a maximum periapsis altitude is achieved. Maximizing the periapsis altitude thus minimizes the post-aerocapture ΔV required to place the vehicle into the desired target orbit. For each given entry flight path angle, an exact point to switch from lift up to lift down can be identified. The closer the vehicle is to the shallow side of the corridor, the closer the switch time is to the entry condition. For an entry at the opposite (steep) side of the corridor, no switch is performed, allowing the vehicle to fly full lift up through the entire flight. The theoretical optimum performance is not a practical method of guiding the vehicle through the flight, however it provides valuable information on how the actual flight algorithm compares to a theoretical optimum performance.

The nominal guidance performance was achieved by adjusting the algorithm initialization parameters for the reference mission. Trajectory simulations with HYPAS show that the algorithm performs well under the given conditions. Important to note however is that for the nominal entry flight path angle (center of corridor) there are periods during the flight in which the bank angle is saturated, full lift up or down. This is attributed to the high ballistic number and high $M/C_L A$ of the vehicle. The lack of control authority of the vehicle drives the guidance to initially fly a very steep profile in order to achieve a high dynamic pressure and thus attain the desired control. Then the vehicle must fly full lift up to maintain an equilibrium glide condition and initiate the exit. When the algorithm was tuned in such a manner to minimize this saturation period, the large random density perturbations overwhelmed the system and causes larger then desired error in apoapsis targeting because the system would not respond aggressively enough. The nominal profile can be seen in Figures 8-13.

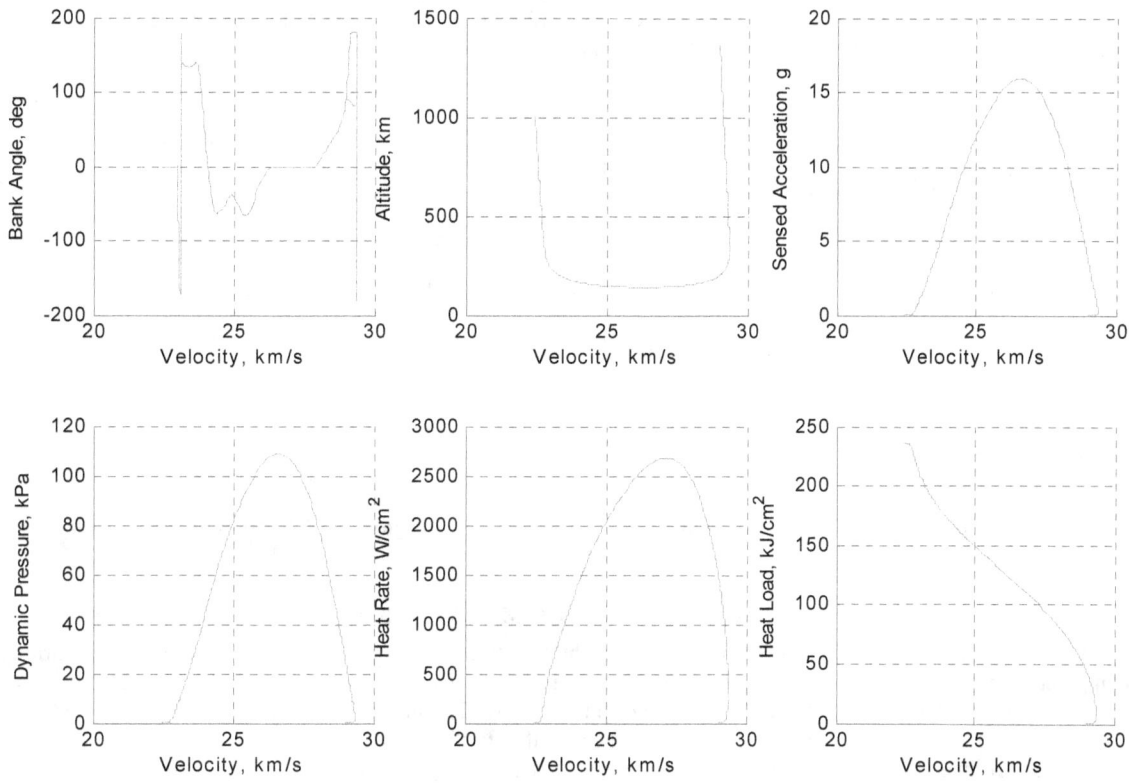

Figures 8-13. Nominal Guidance Profile

AEROCAPTURE GUIDANCE PERFORMANCE

The aerocapture guidance performance was evaluated throughout the entry corridor for nominal and dispersed atmospheres and aerodynamic characteristics. This performance is illustrated in Figure 14 for nominal flight conditions and in figures 15 and 16 for dispersed atmospheres and aerodynamic characteristics. Figure 14 shows apoapsis altitude throughout the entry corridor and shows that the guidance performs well and captures 93% of the corridor, as defined by the 390,000 and 490,000 km apoapsis altitude band. It is important to note that 100% of the cases remained captured into orbit. The results were generated using the nominal mean atmospheric profile (FMINMAX of 0.0), and no random density perturbations were used.

The sensitivity of the HYPAS algorithm to uncertainties in atmospheric conditions and aerodynamic coefficients is illustrated in Figures 15 and 16. The algorithm was subjected to the limits of the expected atmospheric model, the minimum and maximum profile. Figure 15 shows that the guidance handles the maximum profile in a good fashion, only losing slight corridor width on the steep side. Conversely under the minimum atmospheric profile the system the guidance is unable to accommodate the shallower angles. Since the atmosphere has a lower density, it effectively shifts the corridor towards steeper entry angles. Note that in this figure, the shallowest angle, corresponding to −11.9 deg, results in an atmospheric skip, or a negative apoapsis altitude, due to the hyperbolic nature of the orbit.

To simulate possible aerodynamic errors, a ±10 percent variation in lift and drag coefficient was implemented and individually evaluated to assess the extreme response of the

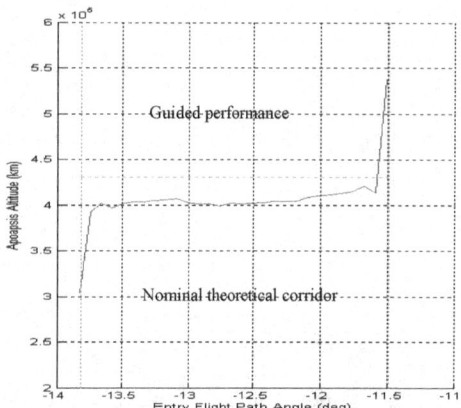

Figure 14. Nominal Guidance Performance

guidance algorithm. The results of this assessment can be seen in Figure 16. Here it can be seen that there is a slight loss of the corridor capture in the +10 percent in C_D and –10 percent in C_L cases, but all cases remain in orbit. The effects are limited to the ability of targeting the desired apoapsis altitude.

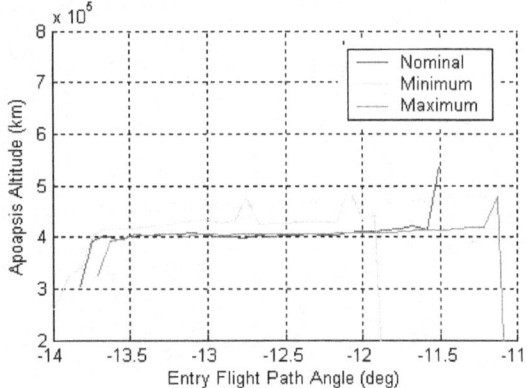

Figure 15. Atmospheric Sensitivity

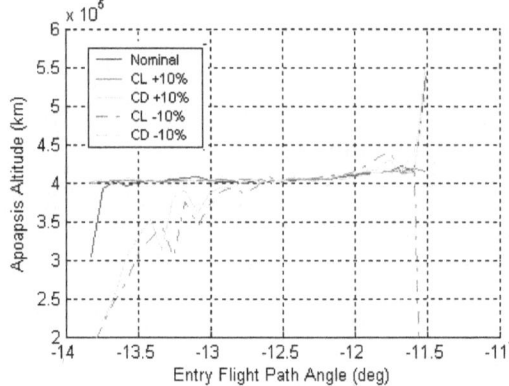

Figure 16. Aerodynamic Sensitivity

ANGLE OF ATTACK MODULATION

Angle-of-attack modulation is used by the Space Shuttle Orbiter entry guidance[10] to improve the performance compared to control with only bank angle modulation. This results in a rapid response to trajectory perturbations by directly changing the area exposed to the flow and thus the drag. The original derivation of HYPAS includes only bank angle modulation, which by moving the lift vector affects the altitude rate of the vehicle and controls the density environment of flight. This then can be translated to drag deceleration modulation. The new set of difficulties encountered in the Neptune mission, lead to a desire for improved control in order to improve the final orbit apoapsis targeting.

The HYPAS algorithm was modified, as shown in equations (3) through (6) to incorporate angle-of attack modulation. The approach taken here is derived from that used for angle of attack modulation on the Space Shuttle. The $\Delta\alpha$ formulation is used to command a change in angle-of-attack relative to the current angle-of-attack. The variation in angle-of-attack from the nominal profile is limited to a specific value, here tested at 3, 5, and 10 deg to limit the effects on aerodynamic heating.

$$\alpha_{cmd} = \alpha + \Delta\alpha \tag{3}$$

$$\Delta\alpha = \frac{C_D(D_{ref} - D)}{\bar{q}G_{\Delta\alpha}} \tag{4}$$

The HYPAS equation then for bank angle is modified to drive the angle-of-attack back to a nominal profile. The original and new bank command equation can be seen in equation (5) and (6),

$$\left(\frac{L}{D}\right)\cos\phi_{cmd} = \frac{C_L}{C_D}\left[\cos\phi_{eq.gl.} - G_h\left(\frac{\dot{h} - \dot{h}_{ref}}{\bar{q}}\right) + G_D\left(\frac{D - D_{ref}}{\bar{q}}\right)\right] \tag{5}$$

$$\left(\frac{L}{D}\right)\cos\phi_{cmd} = \frac{C_L}{C_D}\left[\cos\phi_{eq.gl.} - G_h\left(\frac{\dot{h} - \dot{h}_{ref}}{\bar{q}}\right) + G_D\left(\frac{D - D_{ref}}{\bar{q}}\right) + G_\alpha(\alpha_{nom} - \alpha_{cmd})\right] \tag{6}$$

where ϕ_{cmd} is the commanded bank angle, G_h, G_D, and G_α are the feedback gains for altitude rate, drag deceleration, and angle of attack respectively.

Various ranges of angle of attack modulation limit were initially investigated as an option to improving the performance. In the Monte Carlo performance analysis (Figures 17-24), it can be seen that an angle modulation of ±5 degree can provide significant improvements in the precision of the algorithm. The in-plane ΔV to maneuver the vehicle into the target orbit is also shown. This analysis was conducted as a preliminary assessment. The actual implementation method for angle of attack modulation, such as body flaps, jets, or CG shift, was not investigated, and should be included in further analysis. The addition of angle attack modulation, though increasing the complexity of the system has potential propellant mass savings. Using the 3σ ΔV value for no angle of attack modulation of 378.7 m/s, an initial vehicle mass of 2200 kg, and assuming a typical bipropellant Isp of 320 sec, yields a propellant mass of approximately 250 kg. In contrast, using the 3σ ΔV value for ±10 deg angle of attack modulation of 176.5 m/s, yields a propellant mass of approximately 120 kg, more then 50% savings in mass.

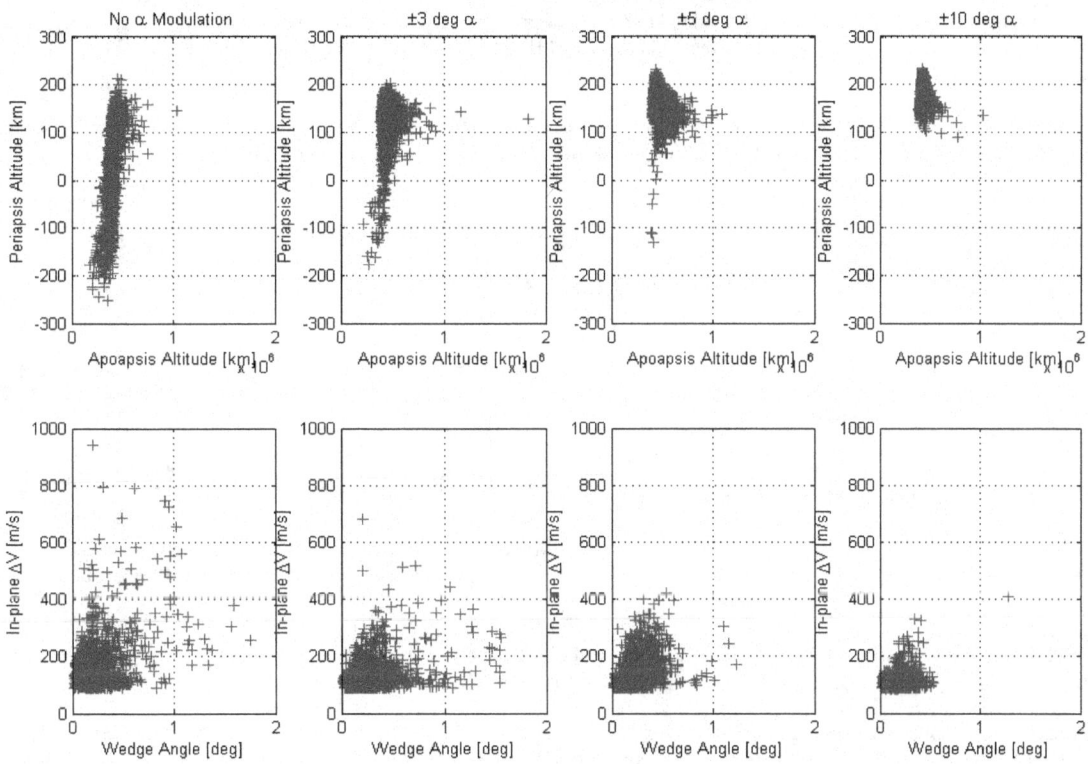

Figures 17-24. Angle of Attack Modulation

SUMMARY AND CONCLUSIONS

The HYPAS guidance algorithm performance for a Neptune aerocapture mission has been investigated. The investigation included development of a nominal guided flight profile, determination of the amount of the theoretical entry corridor captured with the algorithm, and algorithm performance under perturbations in aerodynamics, atmosphere density, and entry state vector using Monte Carlo simulations. The results show that the algorithm performance meets the given propulsive ΔV allotment and satisfies the reference mission requirements. Detailed aerocapture orbit insertion performance for the Neptune Triton Exploration mission can be found in Reference 11.

It was found that mission parameter values unique to the Neptune aerocapture mission investigated here, specifically, the combination of high vehicle ballistic coefficient, high entry velocity, and high energy target orbit, posed difficulties that have not been experienced in previous assessments of aerocapture guidance algorithm performance. The net effect of the extreme values of these parameters is a decrease in the accuracy of achieving the target apoapsis with aerocapture using bank angle control only, thus resulting in a larger than expected post-aerocapture propulsive maneuver. Angle of attack modulation has been incorporated into the HYPAS algorithm to address this issue. Monte Carlo trajectory simulations show that with angle of attack modulation, the accuracy for the given Neptune mission is significantly improved, resulting in a decrease in the propulsive ΔV requirements and reduction in risk of the aerocapture maneuver.

The difficulties encountered in the guidance design for the stressful conditions of the Neptune mission highlights the need to further investigate future changes to the algorithm for this application. This includes an improved reference drag profile to more closely match the flight profile and further investigation of angle of attack control. Further iterations of tuning can also yield an improved bank angle profile, with less control saturation.

REFERENCES

[1]Bailey, R. W., Hall, J. L., Spilker, T. R., and O'kongo, N., Neptune Aerocapture Mission and Spacecraft Design Overview", AIAA 2004-3842, Conference Proceedings of the 40th AIAA/ASME/SAE/ASEE Joint Propulsion Conference and Exhibit, Huntsville, AL, July 2004

[2]Noca, M. A., and Bailey, R. W., "Mission Trades for Aerocapture at Neptune", AIAA-2004-3843, Conference Proceedings of the 40th AIAA/ASME/SAE/ASEE Joint Propulsion Conference and Exhibit, Huntsville, AL, July 2004

[3]Lockwood, M. K., ."Neptune Aerocapture Systems Analysis", AIAA-2004-4951 Conference proceedings of the AIAA Atmospheric Flight Mechanics Conference and Exhibit, Providence, Rhode Island, August 2004.

[4]C. J. Cerimele, and J.D. Gamble, "A Simplified Guidance Algorithm for Lifting Aeroassist Orbital Transfer Vehicles," AIAA-85-0348, AIAA 23rd Aerospace Sciences Meeting, Reno, NV, January 1985.

[5]Edquist, K. T., Hoffman, D. A., Rea, J.R., "Configuration, Aerodynamics and Stability Analysis for a Neptune Aerocapture Orbiter", AIAA-2004-4953, Conference Proceedings of the AIAA Atmospheric Flight Mechanics Conference and Exhibit, Providence, Rhode Island, August 2004

[6]Lockheed Engineering & Sciences Company "User's Guide for the Simulation and Optimization of Rocket Trajectories (SORT) Program Version 8, LESC-30330," NASA Contract NAS 9-17900, October 1998.

[7]Justus, C. G., Duvall, A., Keller, K. W., "Atmospheric Models for Aerocapture Systems Studies", AIAA-2004-4952, Conference Proceedings of the AIAA Atmospheric Flight Mechanics Conference and Exhibit, Providence, Rhode Island, August 2004

[8]Sutton, K., Graves, R. A., Jr., "A General Stagnation Point Convective Heating Equation for Arbitrary Gas Mixtures," NASA TR R-376, NASA Langley Research Center, Hampton VA, Nov. 1971.

[9]Masciarelli, J., Queen, E., "Guidance Algorithms for Aerocapture at Titan," AIAA-2003-4804, Conference Proceedings of the 39th AIAA/ASME/SAE/ASEE Joint Propulsion Conference, Huntsville, AL, July 2003.

[10]Graves, C., Harpold, J., "Shuttle Entry Guidance," AAS 78-147, American Astronautical Society 25th Anniversary Conference, Houston, TX, Oct. 30 – Nov. 2, 1978.

[11]Starr, B.R., Powell, R.W., Westhelle, C.H., Edquist, K.T., and Masciarelli, J.P., "Aerocapture Performance Analysis for a Neptune-Triton Exploration Mission," AIAA-2004-4955, Conference Proceedings of the AIAA Atmospheric Flight Mechanics Conference and Exhibit, Providence, Rhode Island, August 2004.

PRELIMINARY CONVECTIVE-RADIATIVE HEATING ENVIRONMENTS FOR A NEPTUNE AEROCAPTURE MISSION

Brian R. Hollis
NASA Langley Research Center, Hampton, VA 23681

Michael J. Wright and Joseph Olejniczak
NASA Ames Research Center, Moffett Field, CA 94035

Naruhisa Takashima
AMA Inc., Hampton, VA 23666

Kenneth Sutton
National Institute of Aerospace, Hampton, VA 23666

Dinesh Prabhu
ELORET Corp., Sunnyvale, CA 94087

Convective and radiative heating environments have been computed for a three-dimensional ellipsled configuration which would perform an aerocapture maneuver at Neptune. This work was performed as part of a one-year Neptune aerocapture spacecraft systems study that also included analyses of trajectories, atmospheric modeling, aerodynamics, structural design, and other disciplines. Complementary heating analyses were conducted by separate teams using independent sets of aerothermodynamic modeling tools (i.e. Navier-Stokes and radiation transport codes). Environments were generated for a large 5.50 m length ellipsled and a small 2.88 m length ellipsled. Radiative heating was found to contribute up to 80% of the total heating rate at the ellipsled nose depending on the trajectory point. Good agreement between convective heating predictions from the two Navier-Stokes solvers was obtained. However, the radiation analysis revealed several uncertainties in the computational models employed in both sets of codes, as well as large differences between the predicted radiative heating rates.

NOMENCLATURE

A = reference area (m^2)
C = coefficient in reaction rate equation
C_D = drag coefficient
k_f = forward reaction rate (cm^3/mole/s)
L/D = lift-to-drag ratio
m = mass (kg)
n = exponent in reaction rate equation
q = heat-transfer rate (W/cm^2)
t = time (s)
T = temperature (K)
T_a = reaction temperature (K)
T_v = vibrational temperature (K)
Z = axial distance measured from nose (m)
α = angle-of-attack (deg)
β = ballistic coefficient (kg/m^2)
$\quad \beta = m/(C_D A)$
θ = activation temperature (K)
ρ = density (kg/m^3)
Subscripts
rad = radiative
conv = convective

INTRODUCTION AND BACKGROUND

A one year, multi-disciplinary study of a mission to Neptune in which aerocapture would be used to decelerate into orbit has been conducted. Computational analyses of the convective and radiative aeroheating environments which the vehicle would experience are detailed herein, and results from other disciplines are presented in several companion papers[1-5].

NEPTUNE AEROCAPTURE MISSION CONCEPT

In an aerocapture mission, atmospheric drag is employed in place of a conventional propulsion system to decelerate the vehicle into orbit (Fig. 1). Aerocapture can result in large mass savings in comparison to propulsive deceleration. For this study a reference mission concept[1] was developed for a 2017 launch with a 10 year transit to Neptune of an orbiter designed for a scientific investigation of Neptune and its moon Triton. For the reference mission guidelines, it was determined that the mass savings resulting from an aerocapture maneuver would be necessary to deliver the required payload.

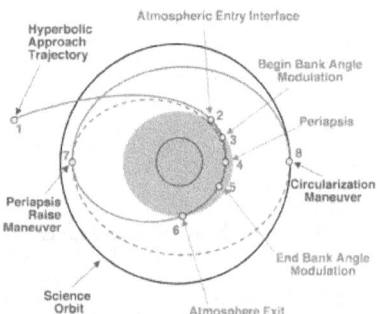

Figure 1: Illustration of Aerocapture Mission

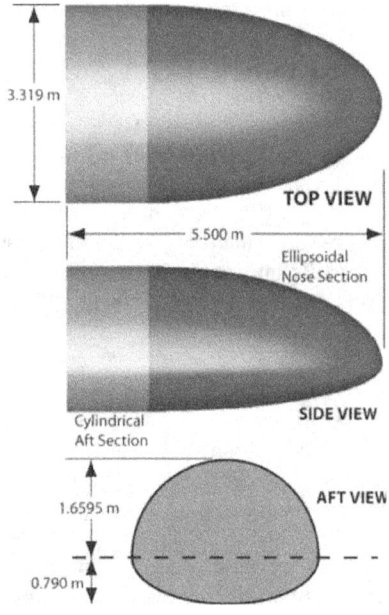

Figure 2: Ellipsled Configuration and Dimensions

VEHICLE CONFIGURATION

An aerodynamic trade-off study[2] was conducted to define the shape of the vehicle's aeroshell. The design objectives were to achieve a lift-to-drag (L/D) ratio of 0.8, minimize the ballistic coefficient (β), maximize the volumetric efficiency, and fit within the launch vehicle shroud. The configuration selected was the "flattened ellipsled" geometry shown in Fig. 2. A basic ellipsled configuration can be defined by an ellipsoid nose section followed by an elliptical cross-section cylinder. The basic ellipsled can then be "flattened" by shrinking the minor axis of the bottom half of the elliptical cross-section of the vehicle.

The dimensions shown in Fig. 2 define the geometry of a 5.50 m long vehicle which was the focus of the first phase of this study. Results from this phase of the study were used to conduct a design iteration, which resulted in a new, smaller vehicle. This second geometry was a scaling of 52.36% from the original design which produced a 2.88 m long vehicle.

VEHICLE TRAJECTORIES

Atmospheric trajectories for both vehicles were generated[3] using the Program for Optimization of Simulated Trajectories[6] with a Neptune atmospheric model[4]. For the large vehicle, a ballistic coefficient (β) of 400 kg/m^2 was used, while for the smaller vehicle a ballistic coefficient of 895 kg/m^2 was used. Reference convective heating rates were computed along these trajectories for a 1.00 m hemisphere using a Sutton-Graves[7] formulation. Several thousand trajectories were simulated, from which worst-cases for convective heat loads and heat rates were identified. These trajectories are shown in Figs. 3 and 4. For the original large vehicle, aeroheating predictions were generated at five points (including peak heating) along the max heat-rate trajectory, while for the final smaller vehicle heating rates were computed at the peak heating point along the max heat-rate trajectory and at seven points along the max heat-load trajectory. These heating predictions were then used to develop the aeroheating environments required to design[5] a Thermal Protection System (TPS) for the vehicle. Free stream conditions for these points are given in Tables 1 and 2. The angle-of-attack for all cases was a α = 40-deg.

108

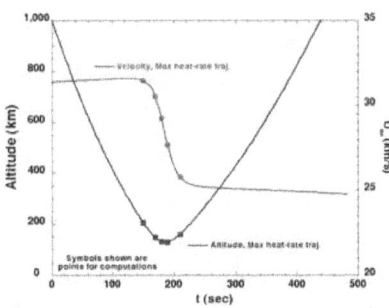

Figure 3: Max Convective Heat-Rate Trajectory for Large Ellipsled

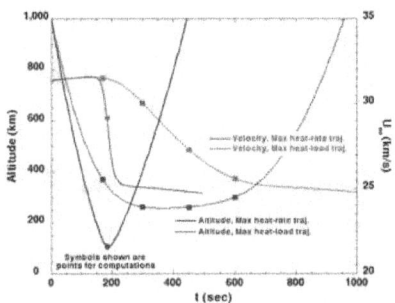

Figure 4: Max Convective Heat-Rate and Max Heat-Load Trajectories for Small Ellipsled

COMPUTATIONAL TOOLS

Convective-radiative heating environments were generated by two independent teams using separate flow field and radiation transport codes. Comparisons of results from the two analyses were performed for each set of trajectory computations in order to verify the results and identify sources of uncertainty.

Within the context of this one-year study, the focus of the work was on the generation of environments for the system study rather than computational tool development. However, computation of the Neptune aerocapture environment posed several challenges which will be discussed in later sections, and further research into computational tools and models for the Neptune missions will be required to address these issues.

FLOW FIELD SOLVERS

Flow field solutions were generated using the LAURA (Langley Aerothermodynamic Upwind Relaxation Algorithm)[8,9] and DPLR[10] (Data Parallel Line Relaxation) codes. Both codes are three-dimensional, structured, finite-volume Navier-Stokes solvers and support multiple-block computations on distributed nodes using the Message Passing Interface (MPI). Inviscid fluxes are computed in LAURA using the Roe flux splitting[11] method with Harten's entropy fix[12] and Yee's symmetric total-variation diminishing limiting[13], while a modified Steger-Warming flux vector splitting[14] with MUSCL extrapolation to third order with a minmod limiter[15] is implemented in DPLR. Previous studies have shown that both codes produce similar results when the same kinetic and transport properties models are implemented in each code (e.g. Refs. 16-17).

A 13-species (H_2, H, H+, He, He+ CH_4, CH_3, CH_2, CH, C_2, C, C+, e-) Neptune atmospheric model was employed for flow field computations with default free stream mass fractions of 0.6246 for H_2, 0.2909 for He, and 0.0846 for CH_4. A simpler 5-species model (H_2, H, H+, He, e-) with freestream mass fractions of 0.668 for H_2 and 0.332 for He was also employed for some numerical studies. Two reaction sets were used: the first was taken from Nelson[18] (with the addition of certain reactions from Park[19] and Leibowitz[20]); in the second set, hydrogen and helium dissociation and ionization reactions were replaced with those from Leibowtiz[20]. Equations for the forward reaction rates for both sets are listed in Tables 3 and 4. Reverse rates were computed from the definition of the equilibrium constant, which was determined by evaluating the Gibbs free energy from thermodynamic data supplied by McBride[21].

A radiative wall equilibrium temperature boundary condition with a surface emissivity of 0.90 was imposed at the surface. "Super-catalytic" behavior (recombination to free stream mass fractions) was imposed to provide conservative heating estimates.

RADIATION TRANSPORT CODES

Uncoupled radiative transport computations were performed using the flow field solutions as input data. An updated version of the RADEQUIL code[22,23] was used to process LAURA inputs and the NEQAIR96[24] code was used to process DPLR inputs.

RADEQUIL and NEQAIR96 are both used to calculate radiative emission and absorption from input flow field properties with the one-dimensional tangent-slab approximation in which it is assumed that radiation transport takes place only in the direction perpendicular to the surface. The populations of the excited states of the various species are assumed to follow Boltzmann distributions. In RADEQUIL, a line group approximation is used to model atomic line transitions and a smeared band approximation is used to model molecular transitions, whereas in NEQAIR96, line-by-line computations are performed for all atomic and molecular transitions. In this study, transitions in hydrogen atoms due to excitation, bound-free photo-ionization, and free-free transition were considered in both codes. Molecular transitions of H_2 were also included in the RADEQUIL analysis. The transitions occurring in C, C+, and C_2 were modeled, but were found to have negligible contributions due to the low concentrations of these species.

RESULTS AND ANALYSIS

Heating environments computed for the large and small ellipsled configurations are presented in this section. Uncertainties due to flow field and radiation transport solver implementation, vibrational non-equilibrium, atmospheric composition, radiation-flow field coupling, and radiation models are also discussed.

5.50 m, β = 400 kg/m^2 Vehicle

Convective heat transfer computations were performed for five points (Table 1) along the max heat-rate trajectory for the large ellipsled configuration using both LAURA and DPLR. Radiation transport calculations were performed only with NEQAIR96 before the vehicle configuration evolved to the smaller, 2.88 m geometry. Centerline convective heating distributions (LAURA and DPLR results essentially identical) for each trajectory point are shown in Fig. 5 and global convective distributions at the peak heating point (t=180 s) along the trajectory are shown in Fig. 6. Peak convective and radiative heating rates (at the nose) for each trajectory point are shown in Fig. 7. The maximum heating rates on the ellipsled were 3833 W/cm^2 for convective heating and 1302 W/cm^2 for radiative heating at t = 180 s. The highest percentage contribution of radiation to the total heating environment was 48% at t = 150 s.

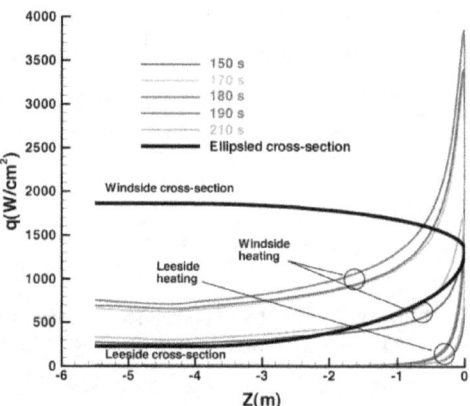

Figure 5: Centerline Convective Heating along Max Heat-Rate Trajectory for Large Ellipsled

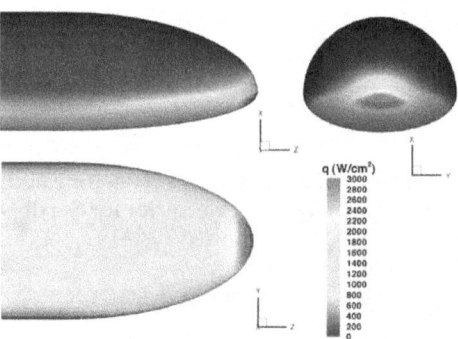

Figure 6: Global Convective Heating Distribution at Peak Heating for Large Ellipsled

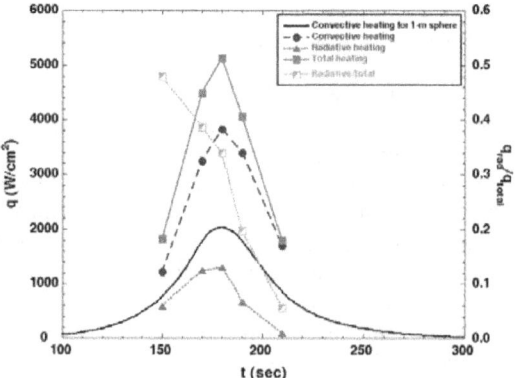

Figure 7: Nose Heating Rates along Max Heat-Rate Trajectory for Large Ellipsled

UNCERTAINTIES IN FLOW FIELD AND RADIATION TRANSPORT METHODS

Analysis of heating results and comparisons between the different flow field and radiation transport codes revealed several areas in which large uncertainties exist in the modeling of the high-energy aerocapture pass through Neptune's atmosphere. Each of these areas will be discussed in the following subsections. In some cases differences were explored using a 0.3 m radius hemisphere geometry (the approximate nose radius of the large ellipsled) in place of the complex, three-dimensional ellipsled geometry, along with the simpler H_2-He atmosphere.

Flow Field Code-to Code Comparisons

Code-to-code comparisons between LAURA and DPLR of convective heating rates resulted in agreement to within ±10% or less, as shown by sample comparisons in Fig. 8 for the large ellipsled and Fig. 9 for the small ellipsled. As a result, where convective heating rates for the ellipsled are shown throughout this paper, they are usually not identified as resulting from either code because the differences were generally very small. However, as will be shown subsequently, differences in kinetic models initially implemented in the two codes which had little influence on convective heating rates had large effects on radiation computations.

Vibrational Nonequilibrium

Stability problems were encountered using both LAURA and DPLR to compute the ellipsled flow fields in the H_2-He-CH_4 Neptune atmosphere. These problems were traced back to the modeling of vibrational non-equilibrium. For the trajectories under consideration, dissociation of the molecular species (H_2, CH_4, CH_3, CH_2, CH, C_2) behind the shock was very rapid. This dissociation left very small concentrations of molecules for which vibrational equilibrium could be defined except in narrow regions at the shock and wall, which made the computational problem very stiff. Examples of this rapid dissociation are given by stagnation-line temperatures and H_2 mole fractions plotted in Fig. 10 for the 0.3 m hemisphere geometry at the peak heating points on the max heat load and max heat rate, $\beta = 895$ kg/m^2 trajectories.

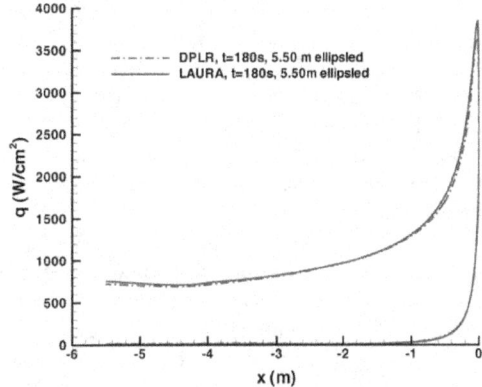

Figure 8: LAURA-DPLR Centerline Convective Heating Comparison for Large Ellipsled

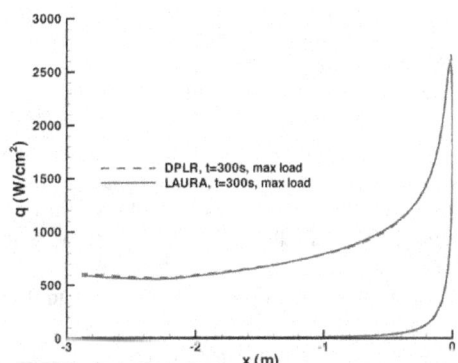

Figure 9: LAURA-DPLR Centerline Convective Heating Comparison for Small Ellipsled

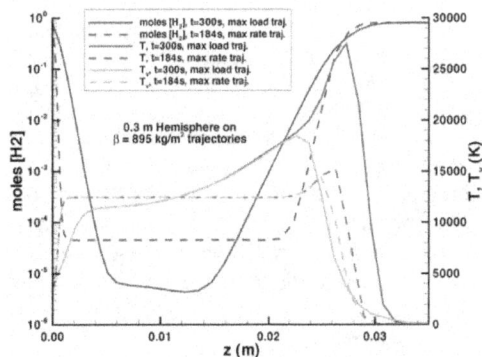

Figure 10: Hemisphere Stagnation-Line Temperature and H_2 Mole Fractions

111

An additional uncertainty in vibrational non-equilibrium computations is the use of Park's two-temperature (TT_v) model[25]. This model was developed for use in air, where the primary diatomic species, N_2 and O_2, have similar vibrational characteristics. The applicability of this model to the Neptune environment is unproven.

The flow field was found to be very nearly in vibrational equilibrium for points along the lower-altitude, max heat-rates trajectories, while vibrational non-equilibrium was present for points along the higher-altitude, max heat-load trajectories. In order to avoid the stability problems which occurred when vibrational non-equilibrium was allowed, all small ellipsled cases were computed with vibrational equilibrium. This approximation had no appreciable affect on convective heating rates, but was recognized as a conservative assumption in the computation of radiative heating rates. The conservatism results from the fact the equilibrium temperatures used to evaluate radiation transport were considerably higher near the shock than the vibrational temperatures which would have been used if vibrational non-equilibrium was allowed. These temperature differences are illustrated for the 0.3 m hemisphere geometry (H_2-He mixture) in Fig. 11.

As an example of the differences resulting from the vibrational equilibrium assumption, the stagnation-point radiative heating levels computed using RADEQUIL with inputs from LAURA for the four cases shown in Fig. 11 were:
Peak convective heat-load trajectory (t = 300 s):
q_{rad} = 6084 W/cm², vibrational non-equilibrium
q_{rad} = 36,950 W/cm², vibrational equilibrium
Peak convective heat-rate trajectory (t = 184 s):
q_{rad} = 5192 W/cm², vibrational non-equilibrium
q_{rad} = 7433 W/cm², vibrational equilibrium
As shown by the above values, the vibrational equilibrium assumption does reasonably well (for radiation calculations) at approximating the radiative heating rates for cases near equilibrium, but for cases in non-equilibrium, this assumption produces extremely conservative results.

The differences in radiative heating levels for the peak heat-load cases were clearly unacceptable; however, the numerical stability problems discussed previously prevented non-equilibrium solutions from being computed on

the ellipsled geometry in a timely manner. Therefore, an approximate correction to the equilibrium radiative heating results was made to obtain order-of-magnitude estimates for non-equilibrium levels. It was found that radiative heating levels approximating those obtained from non-equilibrium flow field computations could be obtained from equilibrium flow field computations if the H_2 molecular transition contributions were neglected. The rationale for this approximation was that H_2 produced the majority of the total radiation in the peak-load trajectory cases with equilibrium modeling, but with non-equilibrium modeling, the H_2 would have been radiating at a much lower temperature. Thus, subtracting the H_2 contribution from the equilibrium total radiative heating rate resulted in radiative heating estimates of the same order of magnitude as non-equilibrium predictions. For the hemisphere cases discussed above, the approximate (H_2 contribution removed) equilibrium radiative heating rates were:
Peak load trajectory (t = 300s):
q_{rad} = 5684 W/cm² equilibrium with H_2 removed
Peak rate trajectory (t = 184 s):
q_{rad} = 4797 W/cm² equilibrium with H_2 removed

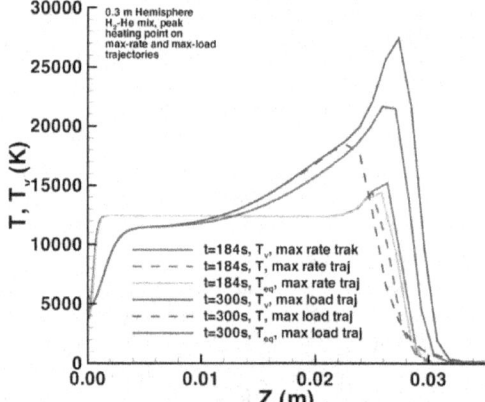

Figure 11: Hemisphere Stagnation Line Translational and Vibrational Temperatures

Radiation Code-to-Code Comparisons
Comparison of LAURA-RADEQUIL and DPLR-NEQAIR96 flow field equilibrium radiative heating levels (with the H_2 contributions removed) were performed for several cases.

For the hemisphere cases discussed in the previous section, the predicted levels were:
Peak convective heat-load trajectory (t = 300 s):
q_{rad} = 5684 W/cm² LAURA-RADEQUIL
q_{rad} = 1800 W/cm² DPLR-NEQAIR96
Peak convective heat-rate trajectory (t = 184 s):

q_{rad} = 4797 W/cm^2 LAURA-RADEQUIL
q_{rad} = 4100 W/cm^2 DPLR-NEQAIR96

For the small ellipsled, the peak (at the nose) radiative heating levels were:
Peak convective heat-load trajectory (t = 300 s):
q_{rad} = 8120 W/cm^2 LAURA-RADEQUIL
q_{rad} = 2200 W/cm^2 DPLR-NEQAIR96
Peak convective heat-rate trajectory (t = 184 s):
q_{rad} = 5610 W/cm^2 LAURA-RADEQUIL
q_{rad} = 4400 W/cm^2 DPLR-NEQAIR96

From these numbers, the differences between the two sets of codes were found to be ~25% for near-equilibrium conditions (along the peak heat-rate trajectory) but were up to ~250% for non-equilibrium conditions (along the peak heat-load trajectory).

As flow field code-to-code comparisons revealed only minor differences (when the same kinetic models were employed), the differences in radiative heating levels were attributed almost entirely to the radiation transport solvers. These differences remained unresolved within the time-frame of this study, but several different assumptions in the radiation transport models of the two codes were noted: RADEQUIL includes more H atomic line transitions (Lyman-$\alpha,\beta,\gamma,\delta,\varepsilon$ Balmer-$\alpha,\beta,\gamma,\delta$, and Paschen-$\alpha,\beta,\gamma$) than NEQAIR96 (Lyman-$\alpha,\beta,\gamma$ Balmer-α,β,γ); NEQAIR96 includes line-by-line calculations of all radiation wavelengths, while RADEQUIL uses a smeared molecular band model; RADEQUIL includes more bound-free photo-ionization transitions (Lyman, Balmer, Paschen, Brackett and approximate integration thereafter to ×) than NEQAIR96 (Lyman and Balmer).

CHEMICAL KINETICS

Comparisons of computations for the large ellipsled showed that the two flow field solvers produced different results for shock stand-off distances and post-shock temperatures. While these differences had very little effect on the convective heating levels, they did lead to different predictions for radiative heating levels. It was determined that the differences were due to the use of the Nelson kinetics in the LAURA computations and the Leibowitz kinetics in DPLR. These differences are illustrated in Figs. 12-13. The results shown are for the 0.3 m hemisphere using LAURA with both the Nelson or Leibowitz kinetics and an H$_2$-He mixture. In Fig. 12, the stagnation line temperatures are shown along with the mole fraction of ionized hydrogen, the ionization rate of which was the main reason for the different temperatures

predicted using the two kinetic models. As shown in Fig. 13, the different models led to only about a ±10% difference in convective heating. While the accuracy of both models for Neptune flow fields needs to be further explored, the Leibowitz kinetics led to much higher radiative heating predictions. Therefore, this more conservative model was employed in both codes for subsequent computations on the small ellipsled.

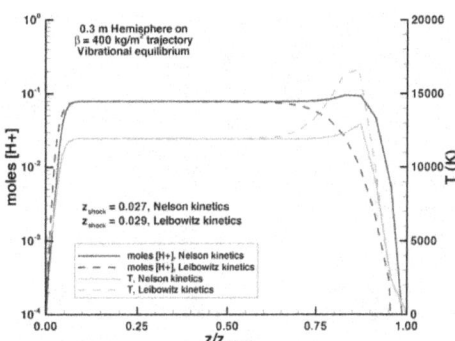

Figure 12: Kinetic Model Effects on Hemisphere Stagnation-Line Temperature and H+ Mole Fractions

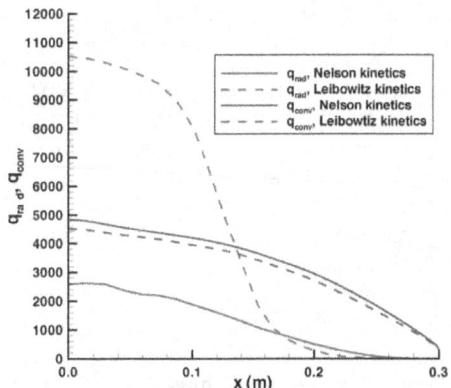

Figure 13: Kinetic Model Effects on Hemisphere Convective and Radiative Heating Distributions

ATMOSPHERIC COMPOSITION

In this study, the baseline atmospheric composition of Neptune was assumed to 0.6246 for H$_2$, 0.2909 for He and 0.0846 for CH$_4$ by mass. However, there is evidence to suggest a trace amount of N$_2$ in Neptune's atmosphere. In order to determine the effects of composition on the heating environment, DPLR-NEQAIR96 computations were performed for the large ellipsled at the peak heating point (t = 180 s) for 5 species (H$_2$/He and products), 11 species

(H_2/He/CH_4 and products without C_2 or He+), and 19 species (H_2/He/CH_4/N_2 and products with different N_2 fractions) compositions. Centerline convective heating distributions and stagnation point convective and radiative heating rates for these cases are shown in Fig. 14. Convective heating levels were found to be relatively insensitive to composition, while radiative heating levels were sensitive to the presence of N_2 through the formation of radiating CN.

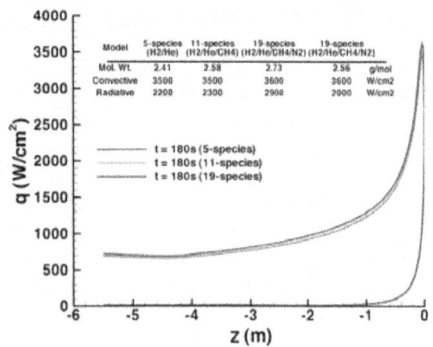

Figure 14: Effects of Atmospheric Composition on Convective and Radiative Heating

Radiation-Flow Field Coupling

In this study, radiation and flow field computations were uncoupled. However, previous studies (e.g. 26) have shown that coupling of computations (feeding the radiation transport results back into the flow field code) can result in significant reductions to the predicted radiative heating levels through non-adiabatic radiative cooling of the flow field. The current uncoupled approach is thus recognized to yield conservative results.

2.88 m, β = 895 kg/m^2 Vehicle

Convective and radiative heat transfer computations were performed for seven points (Table 2) along the max heat-load trajectory and at the peak heating point on the max heat-rate trajectory. As per the discussion in the previous section, the flow field was modeled as being in vibrational equilibrium, while approximate radiative heating rates were computed by neglecting the H_2 contribution. Centerline convective heating distributions for each trajectory point are shown in Figs. 15 and 16. Peak (at the nose) convective and radiative heating rates for each trajectory point are shown in Fig. 17. The maximum convective heating rates were 2575 W/cm^2 on the max-load

trajectory and 7915 W/cm^2 on the max-rate trajectory. Considerable differences were again observed in the radiative heat transfer rates from RADEQUIL and NEQAIR96, but both sets of results showed that radiative heating was of the same order-of-magnitude as convective heating. At the peak heating point on the max-heat-load trajectory, DPLR-NEQAIR96 predicted radiation heating levels at ~45% of the total heating while LAURA-RADEQUIL predicted radiative levels at ~80% of the total heating level.

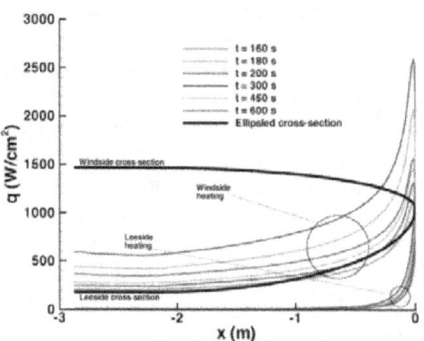

Figure 15: Centerline Convective Heating along Max Heat-Load Trajectory for Small Ellipsled

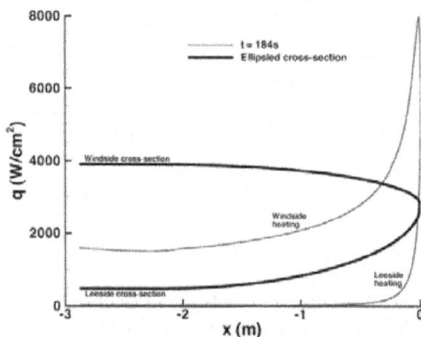

Figure 16: Centerline Convective Heating along Max Heat-Rate Trajectory for Small Ellipsled

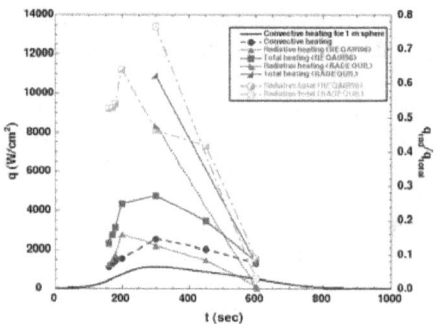

Figure 17: Nose Heating Rates along Max Heat-Rate Trajectory for Small Ellipsled

TPS DEVELOPMENT

Approximate heating rate and integrated heat-load environments were generated from these results for use in TPS material selection and thickness sizing. The convective and radiative rates from the computations were used as anchor points from which to scale the heating time-history outputs from POST for a 1-m hemisphere to the ellipsled geometry. Three environments were generated: a "Low" environment based on the convective heating plus one-half of the radiative heating rate to account for radiative cooling effects, a "Reference" environment based on the total (convective plus radiative) heating rates generated herein, and a "High" environment based on twice the total heating rates to account for turbulent heating augmentation. TPS development based on these environments is discussed in the companion paper by Laub[5].

SUMMARY

Preliminary convective and radiative heating environments for a Neptune aerocapture mission have been computed. Environments were generated both for a large 5.50 m ellipsled and a small 2.88 m ellipsled. Radiative heating constituted up to 80% of the total heating along the trajectories studied.

Because of the expected computational difficulties for this high-velocity aerocapture mission in Neptune's H_2-He-CH_4 atmosphere, heating environments were generated in tandem using LAURA with RADEQUIL and DPLR with NEQAIR96 to compute the flow field and radiation transport properties. This approach was designed to reduce uncertainties and to identify areas in which further research and development of numerical models and tools will be required in order to provide higher confidence in analyses for this class of mission.

The computations were found to agree well for flow field properties and convective heating distributions (when the same kinetic models were employed), but several sources of large uncertainty were identified in the computation of radiative heating.

Kinetic modeling of reactions in the H_2-He-CH_4 Neptune atmosphere was one of the problem areas identified. The use of different reaction sets cited in the literature produced large differences in post-shock stand-off distance and

temperatures, which led to large differences in predicted radiative heating rates.

Vibrational non-equilibrium modeling also presented difficulties. Numerical stability could not be achieved for several cases when a two-temperature, vibrational non-equilibrium model was employed. In order to resolve this issue, vibrational equilibrium was imposed, which was shown to lead to over-prediction of radiative heating levels. Furthermore, even if stability had been achieved, the use of a two-temperature model developed for Earth's N_2-O_2 atmosphere is unproven in Neptune's H_2-He-CH_4 atmosphere.

It was also found that large differences existed in radiative heating rates produced by the two radiation transport codes. While several differences in the physical models incorporated in these codes were identified, the specific reasons for these differences were not identified.

Within the limits of this short-term study, preliminary convective-radiative heating environments for thermal protection system sizing were generated which were sufficient to support a moderate-fidelity vehicle design for a Neptune aerocapture mission. However, in order to complete a high-fidelity design, further development of computational tools and methods for the Neptune environment will be required.

ACKNOWLEDGEMENTS

This research was supported by the In-Space Propulsion Program Office at the NASA Marshall Space Flight Center. Dinesh Prabhu was supported by NASA Ames Research Center grant NAS2-99092 to ELORET Corp. Naruhisa Takashima was supported by NASA Langley Research Center through contract NAS1-00135 to AMA, Inc. Kenneth Sutton was supported through NASA Langley contract NCC1-02043 to the National Institute of Aerospace.

REFERENCES

[1]Lockwood, M. K., "Neptune Aerocapture Systems Analysis", AIAA Paper 2004-4951, Aug. 2004.

[2]Edquist, K. T, Prabhu, R., Hoffman, D., and Rea, J., "Configuration, Aerodynamics, and Stability Analysis for a Neptune Aerocapture Orbiter," AIAA Paper 2004-4953, Aug. 2004.

[3] Starr, B., Powell, R., Westhelle, C., and Masciarelli, J., "Aerocapture Performance Analysis for a Neptune-Triton Exploration Mission," AIAA Paper 2004-4955, Aug. 2004.

[4]Justus, C, Duvall, A., and Keller, V., "Atmospheric Models for Aerocapture Systems Study," AIAA Paper 2004-4952, Aug. 2004.

[5]Laub, B. and Chen, Y., "Challenges for Neptune Aerocapture," AIAA Paper 2004-5178, Aug. 2004.

[6]Powell, R.W., Striepe, S.A., Desai, P.N., Queen, E.M.; Tartabini, P.V., Brauer, G.L., Cornick, D.E., Olson, D.W., Petersen, F.M., Stevenson, R., Engel, M.C.; Marsh, S.M., "Program to Optimize Simulated Trajectories (POST II), Vol. II Utilization Manual," Version 1.1.1.G, May 2000, NASA Langley Research Center, Hampton, VA.

[7]Sutton, K. and Graves, R. A., "A General Stagnation-Point Convective-Heating Equation for Arbitrary Gas Mixtures," NASA TR-R-376, Nov. 1971.

[8]Gnoffo, P. A., "An Upwind-Biased, Point-Implicit Algorithm for Viscous, Compressible Perfect-Gas Flows," NASA TP-2953, Feb. 1990.

[9]Cheatwood, F. M., and Gnoffo, P. A., "User's Manual for the Langley Aerothermodynamic Upwind Relaxation Algorithm (LAURA)," NASA TM 4674, April, 1996.

[10]Wright, M.J., Candler, G.V., and Bose, D., "Data-Parallel Line Relaxation Method for the Navier-Stokes Equations," AIAA Journal, Vol. 36, No. 9, 1998, pp. 1603-1609.

[11]Roe, P. L., "Approximate Riemann Solvers, Parameter Vectors and Difference Schemes," Journal of Computational Physics, Vol. 43, No. 2, 1981, pp. 357-372.

[12]Harten, A., "High Resolution Schemes for Hyperbolic Conservation Laws," Journal of Computational Physics, Vol. 49, No. 3, 1983, pp. 357-393.

[13]Yee, H. C., "On Symmetric and Upwind TVD Schemes," NASA TM 88325, 1990.

[14]MacCormack, R.W. and Candler, G.V., "The Solution of the Navier-Stokes Equations Using Gauss-Seidel Line Relaxation," Computers and Fluids, Vol. 17, No. 1, 1989, pp. 135-150.

[15]Yee, H.C., "A Class of High-Resolution Explicit and Implicit Shock Capturing Methods," NASA TM 101088, Feb. 1989.

[16]Takashima, N., Hollis, B., Olejniczak, J., Wright, M., and Sutton, K., "Preliminary Aerothermodynamics of Titan Aerocapture Aeroshell," AIAA Paper No. 2003-4952, July 2003.

[17]Olejniczak, J., Wright, M., Prabhu, D., Takashima, N., Hollis, B., Zoby, E. V., and Sutton, K. "An Analysis of the Radiative Heating Environment for Aerocapture at Titan," AIAA Paper 2003-4953, July 2003.

[18]Nelson, H. F., Park, C., and Whiting, E. E., "Titan Atmospheric Composition by Hypervelocity Shock-Layer Analysis," Journal of Thermophysics and Heat Transfer, Vol. 5, No. 2, April-June 1991, pp. 157-165.

[19]Park, C., "Radiation Enhancement by Nonequilibrium During Flight Through the Titan Atmosphere," AIAA Paper 1982-0878, June 1982.

[20]Leibowitz, L. P., and Kuo, T., "Ionizational Nonequilibrium Heating During Outer Planetary Entries," AIAA Journal, Vol. 14, No. 9, Sept. 1976, pp. 1324-1329.

[21]McBride, B. J., Zehe, M. J., and Gordon, S., "NASA Glenn Coefficients for Calculating Thermodynamic Properties of Individual Species," NASA TP 2002-211556, Sept. 2002.

[22]Nicolet, W. E., "User's Manual for the Generalized Radiation Transfer Code (RAD/EQUIL)," NASA-CR-116353, Oct. 1969.

[23]Nicolet, W. E., "Advanced Methods for Calculating Radiation Transport in Ablation-Product Contaminated Boundary Layers," NASA-CR-1656, Sept. 1970.

[24]Whiting, E. E., Yen, L., Arnold, J. O. and Paterson, J. A., "NEQAIR96, Nonequilibrium and Equilibrium Radiative Transport and Spectra Program: User's Manual," NASA RP-1389, Dec. 1996.

[25]Park, C., "Assessment of Two Temperature Kinetic Model for Ionizing Air," AIAA Paper 1987-1574, June 1987.

[26]Wright, M. J., Bose, D., and Olejniczak, J., "The Impact of Flowfield-Radiation Coupling on Aeroheating for Titan Aerocapture," AIAA Paper 2004-0484, Jan. 2004.

Table 1: Free Stream Conditions for Large Ellipsled Trajectory Points

Trajectory	Time (s)	Altitude (m)	Density (kg/m3)	Temperature (K)	Velocity (m/s)
Max convective heat rate	150	207,090	1.319E-05	132.41	31,450
Max convective heat rate	170	148,079	8.392E-05	106.95	30,534
Max convective heat rate	180	132,186	1.450E-04	103.16	29,243
Max convective heat rate	190	130,444	1.538E-04	102.95	27,670
Max convective heat rate	210	160,550	5.397E-05	114.43	25,777

Table 2: Free Stream Conditions for Small Ellipsled Trajectory Points

Trajectory	Time (s)	Altitude (m)	Density (kg/m3)	Temperature (K)	Velocity (m/s)
Max convective heat rate	184	107,702	3.513E-04	96.95	29,158
Max convective heat load	160	389,739	3.670E-06	187.50	31,524
Max convective heat load	170	370,114	5.149E-06	186.60	31,506
Max convective heat load	180	352,568	7.016E-06	185.70	31,478
Max convective heat load	200	323,363	1.239E-05	183.70	31,374
Max convective heat load	300	261,092	3.709E-05	180.01	30,049
Max convective heat load	450	260,262	3.768E-05	179.93	27,309
Max convective heat load	600	299,445	1.822E-05	182.67	25,581

Table 3: Nelson-Park Kinetic Model

#	$K_f = CT_a^n\, e^{-(\theta/Ta)}$	C (cc/mol/s)	n	θ (K)	T_a (K)	Ref.
1	$CH_4 + M \rightarrow CH_3 + H + M$	2.25×10^{27}	-1.87	52,900	$(TT_v)^{0.5}$	18-19
2	$CH_3 + M \rightarrow CH_2 + H + M$	2.25×10^{27}	-1.87	54,470	$(TT_v)^{0.5}$	18-19
3	$CH_2 + M \rightarrow CH + H + M$	2.25×10^{27}	-1.87	50,590	$(TT_v)^{0.5}$	18-19
4	$CH + M \rightarrow C + H + M$	1.13×10^{19}	-1.00	40,193	$(TT_v)^{0.5}$	18-19
5	$C_2 + M \rightarrow C + C + M$	9.68×10^{22}	-2.00	71,000	$(TT_v)^{0.5}$	18-19
6	$H_2 + M \rightarrow H + H + M$	1.47×10^{19}	-1.23	51.950	$(TT_v)^{0.5}$	18-19
7	$H_2 + C \rightarrow CH + H$	1.80×10^{14}	0.00	11,490	T	18-19
8	$C + e^- \rightarrow C^+ + e^- + e^+$	3.90×10^{33}	-3.78	130,000	T_v	18-19
9	$H + e^- \rightarrow H^+ + e^- + e^+$	5.90×10^{37}	-4.00	157,800	T_v	18-19
10	$He + e^- \rightarrow He^+ + e^- + e^+$	1.33×10^{13}	0.50	286,160	T_v	20

Table: Leibowitz-Nelson-Park Kinetic Model

#	$K_f = CT_a^n\, e^{-(\theta/Ta)}$	C (cc/mol/s)	n	θ (K)	T_a (K)	Ref.
1	$CH_4 + M \rightarrow CH_3 + H + M$	2.25×10^{27}	-1.87	52,900	$(TT_v)^{0.5}$	18-19
2	$CH_3 + M \rightarrow CH_2 + H + M$	2.25×10^{27}	-1.87	54,470	$(TT_v)^{0.5}$	18-19
3	$CH_2 + M \rightarrow CH + H + M$	2.25×10^{27}	-1.87	50,590	$(TT_v)^{0.5}$	18-19
4	$CH + M \rightarrow C + H + M$	1.13×10^{19}	-1.00	40,193	$(TT_v)^{0.5}$	18-19
5	$C_2 + M \rightarrow C + C + M$	9.68×10^{22}	-2.00	71,000	$(TT_v)^{0.5}$	18-19
6	$H_2 + M \rightarrow H + H + M$	1.04×10^{19}	-1.00	51.950	$(TT_v)^{0.5}$	20
7	$H_2 + C \rightarrow CH + H$	1.80×10^{14}	0.00	11,490	T	18-19
8	$C + e^- \rightarrow C^+ + e^- + e^+$	3.90×10^{33}	-3.78	130,000	T_v	18-19
9	$H + e^- \rightarrow H^+ + e^- + e^+$	2.28×10^{13}	0.50	157,800	T_v	20
10	$H + e^- \rightarrow H^+ + e^- + e^+$	4.11×10^{13}	0.50	116,100	T_v	20
11	$H + H \rightarrow H^+ + e^- + H$	6.17×10^{10}	0.50	116,100	T	20
12	$H + He \rightarrow H^+ + e^- + He$	4.88×10^{10}	0.50	116,100	T	20
13	$He + e^- \rightarrow H^+ + e^- + e^+$	1.33×10^{13}	0.50	286,160	T_v	20

TPS CHALLENGES FOR NEPTUNE AEROCAPTURE

B. Laub & Y.K. Chen
NASA Ames Research Center, Moffett Field, CA 94035

A study to develop a conceptual design for an aerocapture mission at Neptune was conducted by a NASA systems analysis team comprised of technical experts from several NASA centers. Multidisciplinary analyses demonstrated that aerocapture could be accomplished at Neptune with a rigid aeroshell with a flattened ellipsled geometry flying at a nominal angle-of-attack of 40 degrees entering the Neptune atmosphere at an inertial entry velocity of $\approx$ 29 km/s. Aerothermal analyses demonstrated that both the peak convective and radiative heating rates in the stagnation region are very severe. Furthermore, due to the duration of the aerocapture trajectory, the total integrated heat loads are extremely large. TPS sizing analyses were conducted for a limited range of candidate TPS materials since such high peak heat fluxes limit candidate materials to dense, carbonaceous ablators. On the windward side, in regions away from the stagnation region, lower density ablators may suffice. Low-density ablators also are viable candidates on the lee side. However, there are significant uncertainties associated with the turbulent and radiative heating rates. TPS mass requirements for this mission are very large, and the ablator thickness requirements may be beyond current manufacturing capabilities.

NOMENCLATURE

σ	Standard deviation
ΔV	Change in velocity
A	Vehicle cross-sectional area
ARA	Applied Research Associates
ARC	NASA Ames Research Center
c.g.	Center-of-gravity
C_A	Axial force coefficient
C-C	Carbon-carbon
C_D	Vehicle drag coefficient
CFD	Computational Fluid Dynamics
C_N	Normal force coefficient
CP	Carbon phenolic
DOD	Department of Defense
EJGA	Earth-Jupiter gravity assist
EOL	End-of-life
JSC	NASA Johnson Space Center
L/D	Lift-to-drag ratio
LaRC	NASA Langley Research Center
LMA	Lockheed-Martin Astronautics
M	Vehicle mass
$M/C_D A$	Ballistic coefficient, kg/m^2
SCBA	Strip Collar Bonding Approach
SEP	Solar Electric Propulsion
TPS	Thermal Protection System
VJGA	Venus-Jupiter gravity assist

119

INTRODUCTION

Through detailed trade studies,[1] the mission analysis team determined that aerocapture at Neptune could be accomplished with a rigid aeroshell configured as a flattened ellipsled with a lift-to-drag (L/D) ratio of ≈ 0.80 and a ballistic coefficient (M/C_DA) of ≈ 895 kg/m^2. The configuration is illustrated in Figure 1.

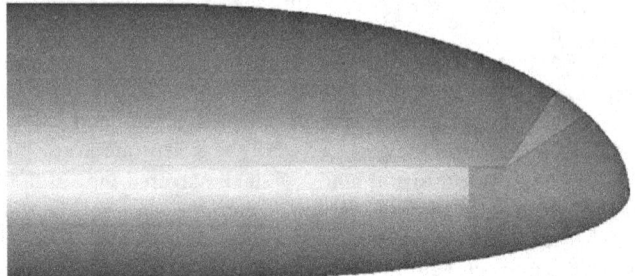

Figure 1. Flattened ellipsled geometry (2.88 m).

A mission analysis study[2] evaluated options to deliver an orbiter and two probes to Neptune. It was concluded that chemical trajectories exist that can deliver the required mass for an orbiter and two probes (~ 2000 kg) using aerocapture at Neptune but would require an Earth-Jupiter gravity-assist (EJGA) and the Delta IV H booster. EJGAs would necessitate an auxiliary chemical propulsion "stage" providing 0.5-3 km/s post launch deep space maneuvers. This study also concluded that trajectories also exist that can deliver the required mass for an orbiter and two probes using solar electric propulsion (SEP) and aerocapture. These options would use a Venus-Jupiter gravity-assist (VJGA) and the Delta IV H. The inertial entry velocity for these trajectories is around 28-30 km/s for 10-11 years flight time. It was concluded that a reasonable mission design (in terms of maximizing delivered mass while minimizing trip time) could be accomplished with a launch on February 17, 2017 on a Delta IV H with a total launch mass of 3298 kg dry (4735 kg wet). The mission would use a 6-engine SEP system (30 kW, EOL) and a VJGA. This would be a 12.25 year Neptune Orbiter Mission with a 10.25-year trip time and 2 (or more) years in Neptune Orbit. The payload would include two identical probes and would use aerocapture to establish a target orbit of 430,000 km x 3,986 km (which would enable Triton observations), with a 157.3 degree inclination, retrograde.

The atmosphere around Neptune is composed primarily of hydrogen and helium with small concentrations of methane and nitrogen. The nominal atmospheric composition is approximately 80% H_2, 19% He, and 1% CH_4. Justus et al[3] developed a model for the Neptune atmosphere (Neptune-GRAM) based on data from Voyager and elsewhere. Allowing for measurement uncertainties and the expected latitudinal, altitudinal, seasonal and diurnal variations, the model accounts for a significant variation in atmospheric density with altitude, as shown in Figure 2.

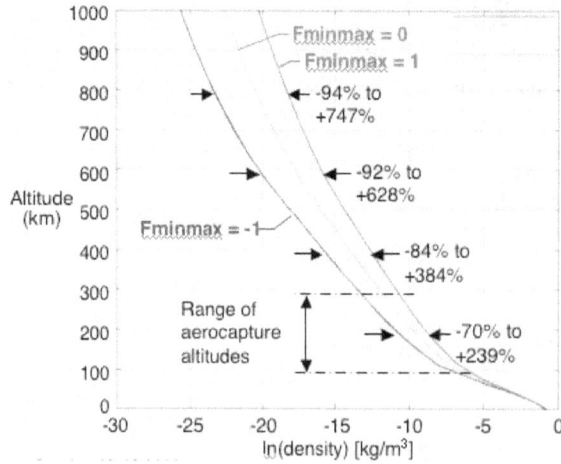

Figure 2. Neptune-GRAM engineering models for the density of the Neptune atmosphere.

Relatively large measurement uncertainties are represented within a prescribed envelope of minimum-average-maximum density versus altitude. A single model input parameter (Fminmax) allows users of Neptune-GRAM to select where within the min-max envelope a particular simulation will fall. Fminmax = -1, 0, or 1 selects minimum, average, or maximum conditions, respectively, with intermediate values determined by interpolation. The percentages shown at selected altitudes in Figure 2 indicate the range of density uncertainty between the minimum density model and the maximum density model in comparison to the average.

An evaluation of the aerodynamic stability[4] of the flattened ellipsled was conducted. The aerodynamic database was developed from viscous LAURA CFD[5,6,7] solutions. A preliminary stability analysis showed that the flat-bottom ellipsled is longitudinally and laterally stable. An assessment of the aerodynamic uncertainties was based on a JSC ellipsled analysis for Mars, consistent with the X-33 aerodynamic database uncertainty model (Ref. NASA TP-1998-206544).

Using these results for vehicle aerodynamics, a Neptune aerocapture simulation and Monte Carlo analysis approach[8] was developed. Three sets of uncertainties were included in the Monte Carlo analysis: delivery errors in initial states, atmospheric variability (mean range and perturbations) and several aerodynamic factors – C_A, C_N, trim angle of attack (pitching moment), c.g. uncertainty. The parameters are each randomly varied over a specified range and distribution. More than 2000 trajectory simulations are completed in one Monte Carlo analysis and the results are used for aerocapture performance statistics to determine robustness, margin and risk, guidance development, stress case identification (control algorithm development to be looked at in the future), and statistical distributions of critical parameters and design trajectories for sub-system design. The analysis demonstrated that 100% of possible aerocapture trajectories successfully achieve aerocapture, as seen in Figure 3. Furthermore, a 360 m/sec 3-σ ΔV would be required to raise the periapsis to 3986 km and correct the apoapsis to 430,000 km. The results were used to select a design trajectory for the evaluation of aerodynamic heating and associated TPS performance. For this purpose the results for heat load vs. heat rate were evaluated, as shown in Figure 4. Monte Carlo trajectory #1647 was selected for the reference concept TPS design since it provides the highest heat load trajectory of the all the considered cases and the heat rate is in the 98th percentile.

Figure 4 illustrates that the Monte Carlo trajectories are in the high heat rate, low heat load range of the lift up, lift down corridor. In this corridor, the peak heating rates[‡] range from a lift up maximum of 3155 W/cm^2 to a lift down maximum of 1122 W/cm^2. Similarly, the heat loads[‡] range from a lift up minimum of 185 kJ/cm^2 to a lift down maximum of 442 kJ/cm^2.

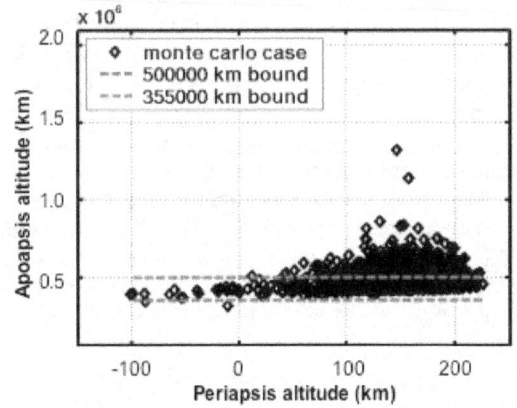
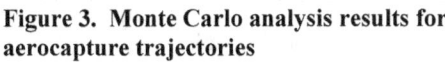
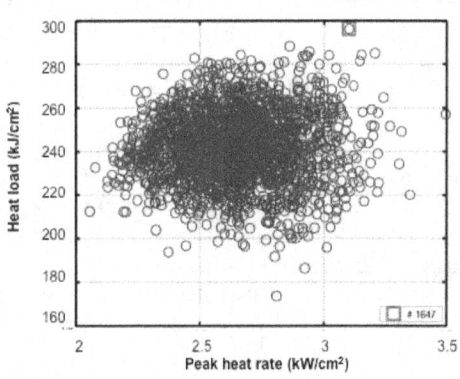

Figure 3. Monte Carlo analysis results for aerocapture trajectories

Figure 4. Heat load vs. heat rate from Monte Carlo results

[‡] The heat rates and heat loads shown are simply used as an aeroheating indicator and actually are for convective stagnation point only and calculated for a 1m-nose radius.

CANDIDATE TPS MATERIALS

Early in the study, NASA Ames did preliminary analyses to estimate the anticipated heating environment for Neptune aerocapture. For that purpose, Ames evaluated the flattened ellipsled geometry on the limiting lift up (undershoot) trajectory with a minimum density atmosphere model. At the time, the preliminary vehicle design had a ballistic coefficient of 400 kg/m^2 in contrast to 895 kg/m^2 for the final design.

Convective heating was calculated through three-dimensional CFD solutions with the DPLR[9] code and radiative heating was calculated with the NEQAIR[10] code using an 11 species model. Figure 5 illustrates heating rates at selected locations on the body at a time in the trajectory when peak heating is experienced. The first numbers are radiative heating rates and the adjoining numbers are convective heating rates. On the basis of these values candidate TPS materials were defined for various regions on the body.

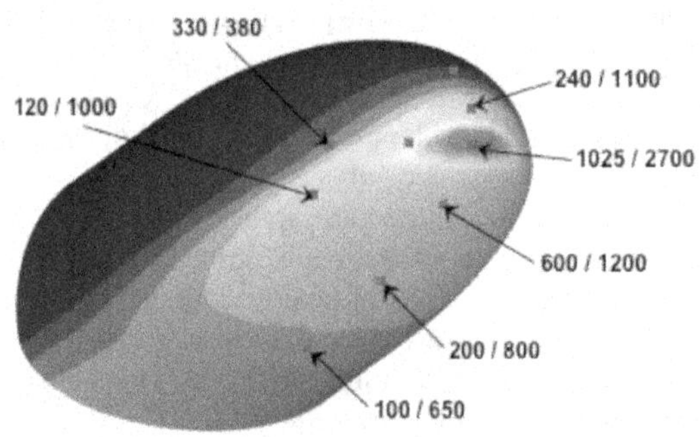

$$\dot{q}_{rad} / \dot{q}_{conv} \left(W/cm^2 \right)$$

Figure 5. Preliminary radiative and convective heating rates on lift up trajectory.

In the windside stagnation region of the vehicle, where heating rates are the highest, candidate materials and their characteristics are summarized in Table 1. Fully dense carbon phenolic is a robust, mature, and well-characterized material that is capable of handling high heat fluxes and pressures. It was used by NASA as the forebody TPS on the Pioneer Venus and Galileo probes and as rocket nozzles for solid propellant boosters. It is well suited for severe heating environments where significant ablation would be anticipated. Its density and composition prevent it being a very good thermal insulator and, as a result, it would be a heavy TPS solution for more modest environments. Reduced density carbon phenolic is a *notional* material based on retaining the good ablation performance associated with a carbon reinforced phenolic composite but improving its thermal insulation performance by reducing the composite density. Such materials were briefly studied in the 1980s and a few composites were fabricated and tested with encouraging results. The Genesis TPS is comprised of a thin carbon-carbon facesheet attached (through co-processing) to low-density carbon fiberform insulation. It was manufactured as a single-piece heatshield for the 60° cone-shaped forebody of the 1.5 m diameter Genesis probe. The material was qualified through arc jet testing to heat fluxes as high as 700 W/cm^2 for the Genesis mission. Because of its all-carbon construction, it is not the best thermal insulator. The modified Genesis TPS is also a *notional* material, in which part of the carbon fiberform insulation is replaced with a low-density ceramic tile to improve its thermal insulation performance.

For the windside expansion region, where heating rates are somewhat lower, candidate materials are summarized in Table 2. Many of the same materials that are candidates for the windside stagnation region are also viable candidates for the windside expansion region. The exception is fully dense carbon phenolic, which has been replaced with PICA (Phenolic Impregnated Carbon Ablator), in which a low-density carbon fiberform is impregnated with phenolic resin. PICA is being used as the forebody TPS on the 60° cone-shaped forebody of the 0.83 m diameter Stardust probe. It has been tested to heat fluxes in excess of 1500 W/cm^2 with good ablation performance, but begins to exhibit char spall at pressures in excess of 0.6 atm.

Table 1. Candidate TPS materials for windside stagnation region

Material	Fully-dense CP	Reduced density CP	Genesis	Modified Genesis
Density (g/cm³)	1.45	0.967	1.8/0.18	1.8/0.18/0.192
Description	Fully-dense tape-wrapped or chopped molded heritage material (used on Pioneer-Venus and Galileo probes).	Reduced density tape-wrapped or chopped-molded composite (studied in the 70s and 80s).	Carbon-carbon facesheet over carbon fiberform insulator.	Carbon-carbon facesheet over carbon fiberform insulator over low-density ceramic insulator (e.g., AETB 20/12).
Optical Properties Solar absorptance Total hemis. emittance	≈ 0.90 ≈ 0.90	≈ 0.90 ≈ 0.90	≈ 0.90 ≈ 0.90	≈ 0.90 ≈ 0.90
Performance limits	Ablative. No oxidation in Neptune's H_2-He environment, but will sublime at high heat fluxes. Will spall at very high heat fluxes (> 25 kW/cm²).	Ablative. No oxidation in Neptune's H_2-He environment, but will sublime at high heat fluxes. Will spall at very high heat fluxes (< 25 kW/cm²) and pressures).	Ablative. No oxidation in Neptune's H_2-He environment, but will sublime at high heat fluxes. Robust ablator, but could spall at very high heat fluxes, depending upon C-C processing.	Ablative. No oxidation in Neptune's H_2-He environment, but will sublime at high heat fluxes Robust ablator, but could spall at very high heat fluxes, depending upon C-C processing.
Uncertainties	A very poor insulator. Best application is for environment where significant ablation is expected. Difficult to fabricate tape-wrapped construction in thicknesses > 2 inches.	Better insulator than fully dense. Spall threshold = $f(\dot{q}, p)$ unknown and (most likely) a function of composite density	Genesis concept only qualified to ≈ 700 W/cm² but should be capable of much higher fluxes. Fabricate as 1-piece or tiles? Issue is (likely) expansion compatibility	Same issues as Genesis with added complexity of bonding a 3rd material.

Finally, for the leeside region, where heating rates are much lower and radiative heating is estimated to be negligible, candidate materials are summarized in Table 3. SLA-561V is a silicone-based ablative in honeycomb developed by Lockheed-Martin (LMA) and is a flight-proven material that was used as the forebody heatshield on the Mars Viking, Mars Pathfinder and Mars Exploration Rover entry probes. The SRAM17 and SRAM20 materials are also silicone-based and were developed by Applied Research Associates (ARA). They are manufactured by a different technique, called the Strip Collar Bonding Approach (SCBA). PhenCarb20 is another ARA material manufactured by the SCBA technique, which uses a phenolic resin rather than silicone. It is capable of handling higher heat fluxes than the silicone-based materials, at the sacrifice of insulation efficiency.

AEROTHERMAL ENVIRONMENTS

A team from NASA's Langley Research Center (LaRC) and NASA's Ames Research Center (ARC) who

Table 2. Candidate TPS materials for windside expansion region

Material	PICA	Reduced density CP	Genesis	Modified Genesis
Density (g/cm³)	0.240	0.967	1.8/0.18	1.8/0.18/0.192
Description	Low-density carbon fiberform partially filled with phenolic resin.	Reduced density tape-wrapped or chopped-molded composite (studied in the 70s and 80s).	Carbon-carbon facesheet over carbon fiberform insulator.	Carbon-carbon facesheet over carbon fiberform insulator over low-density ceramic insulator (e.g., AETB 20/12).
Optical Properties Solar absorptance Total hemis. emittance	≈ 0.90 ≈ 0.80	≈ 0.90 ≈ 0.90	≈ 0.90 ≈ 0.90	≈ 0.90 ≈ 0.90
Performance limits	Ablative. No oxidation in Neptune's H₂-He environment, but will sublime at high heat fluxes. Excellent low-density ablator, but not the best insulator. Will spall at high heat fluxes and pressures.	Ablative. No oxidation in Neptune's H₂-He environment, but will sublime at high heat fluxes. Will spall at very high heat fluxes (< 25 kW/cm²) and pressures).	Ablative. No oxidation in Neptune's H₂-He environment, but will sublime at high heat fluxes. Robust ablator, but could spall at very high heat fluxes depending upon C-C processing.	Ablative. No oxidation in Neptune's H₂-He environment, but will sublime at high heat fluxes Robust ablator, but could spall at very high heat fluxes, depending upon C-C processing.
Uncertainties	Should be better insulator than reduced density CP. Spall threshold = $f(\dot{q}, p)$ unknown.	Better insulator than fully dense. Spall threshold = $f(\dot{q}, p)$ unknown and (most likely) a function of composite density.	Genesis concept only qualified to $\approx$ 700W/cm² but should be capable of much higher fluxes. Fabricate as 1-piece or tiles? Issue is (likely) expansion compatibility.	Same issues as Genesis with added complexity of bonding a 3ʳᵈ material.

conducted independent flow field and radiative transport computations did the aerothermal environment definition of the flattened ellipsled geometry for Neptune aerocapture.[11] The aeroheating team held regular meetings to compare results and identify differences in thermochemical and radiation modeling, assumptions, boundary conditions, etc. Very good agreement was attained for laminar convective heating between LaRC's LAURA[5,6,7] code and ARC's DPLR[9] code. That was not true for turbulent heating where the LAURA code predicted significantly higher turbulent heating rates in comparison to DPLR.[11]

Radiative heating was calculated by LaRC using the RADEQUIL[12,13] code and by ARC using the NEQAIR[10] code. Again, there were significant differences, as discussed in Ref. 11. Based on these results, *estimates* of the heating environments along Monte Carlo trajectory #1647 were completed, and factors ("Low", "Med", "High") were applied to nominal predictions to represent the range of uncertainties. Transition to turbulence due to significant ablation resulting from high total heating rates is expected.

Table 3. Candidate TPS materials for leeside region

Material	SLA-561V	SRAM 17	SRAM 20	PhenCarb-20
Density (g/cm^3)	0.256	0.272	0.320	0.320
Description	Low-density cork silicone composite in Flexcore honeycomb.	Low-density cork silicone composite fabricated w/ strip collar bonding technique.	Low-density cork silicone composite fabricated w/ strip collar bonding technique.	Low-moderate density carbon fiber reinforced phenolic composite fabricated w/strip collar bonding technique.
Optical Properties Solar absorptance Total hemis. emittance	≈ 0.50 ≈ 0.78	≈ 0.50 ≈ 0.78	≈ 0.50 ≈ 0.78	≈ 0.50 ≈ 0.78
Performance limits	Charring Ablator. No (little) recession at heat fluxes ≤ 100 W/cm^2. Char spall at pressures greater than ≈ 0.25 atm. Differential recession between matrix and honeycomb may cause BL transition.	Charring Ablator. No recession due to oxidation, but will sublime at high heat fluxes. Differential recession between composite and interface strips may cause BL transition.	Charring Ablator. No recession due to oxidation, but will sublime at high heat fluxes. Differential recession between composite and interface strips may cause BL transition.	Charring Ablator. No recession due to oxidation, but will sublime at high heat fluxes. Differential recession between composite and interface strips may cause BL transition.
Uncertainties	Data on recession rates at high heat fluxes (> 100 W/cm^2) relatively sparse.	Low density composite will probably spall. Spall threshold = $f(\acute{X}p)$ unknown.	Low density composite will probably spall. Spall threshold = $f(\acute{X}p)$ unknown.	Low density composite will probably spall. Spall threshold = $f(\acute{X}p)$ unknown.

TPS SIZING

For purposes of evaluation, the vehicle was divided into 4 zones for TPS sizing as shown in Figure 6. The candidate TPS materials were selected and sized, with the FIAT[14] code, for the maximum heating point in each zone. The heatshield (forebody) is defined by zone 1 + zone 2. The backshell (aftbody) is defined by zone 3 + zone 4. Post-aerocapture aeroshell separation occurs between the heatshield and backshell.

For the heatshield (Zones 1 and 2), the material models employed with FIAT were extrapolated to heating conditions beyond the range for which they have been validated. The exception is fully dense carbon phenolic that has exhibited excellent performance at even more severe ground and flight conditions in development of the heatshield for the Galileo entry probe. In principle, the other material candidates should perform adequately at these conditions as similar materials have been tested with high-energy lasers at such conditions with good performance.

For the backshell (Zones 3 and 4), the material models employed with FIAT have been validated with ground test data at similar heating conditions and extrapolation was not required.

NOSE REGION (ZONE 1)

The "Low" and "Med" aeroheating rates and loads along Monte Carlo trajectory #1647 are shown in Figure 7. After further aeroheating analyses, it was concluded that the "High" values are outside the expected range. The "Med" level of aeroheating rates and loads was used for TPS sizing for the reference vehicle and is labeled "Reference" in all the following figures.

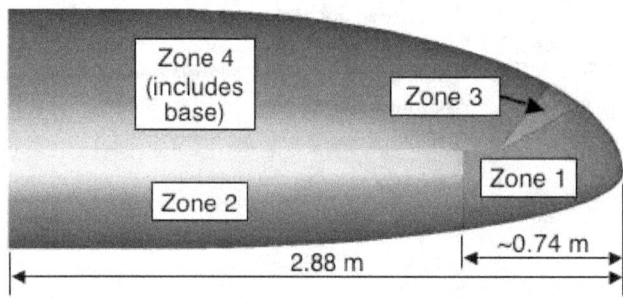

Figure 6. Zones selected for TPS evaluation.

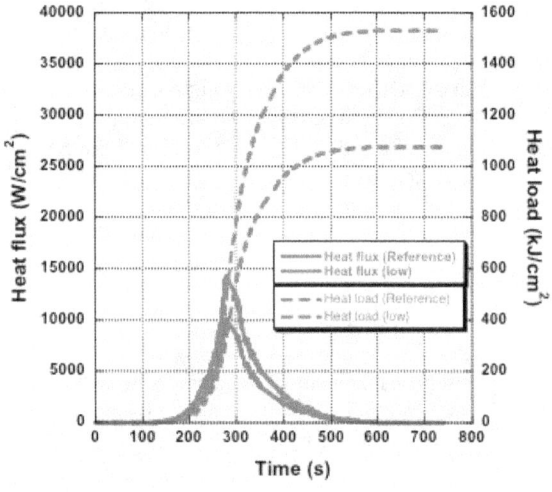

Figure 7. Total low and medium (Reference) heat rates and heat loads for nose region (Zone 1)

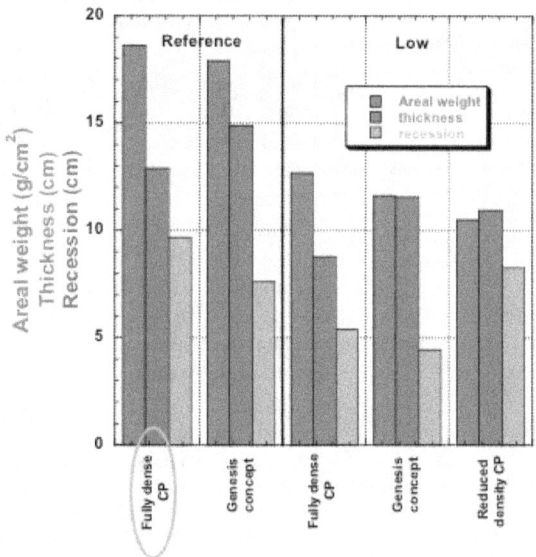

Figure 8. TPS sizing at the nose (Zone 1).

Note that the peak heat fluxes are in the range from 10-15 kW/cm². Only dense, carbonaceous materials can accommodate such high heat fluxes reliably. Furthermore, the total heat loads are in the range from 1076 to 1530 kJ/cm², significantly larger than the heat load for the most severe mission ever flown, i.e., Galileo. Consequently, for the reference ("Med") condition, only fully dense carbon phenolic (ρ = 1.45 g/cm³) and the Genesis concept (carbon-carbon facesheet over carbon fiberform insulator) were considered. As shown in Figure 8, fully dense carbon phenolic (CP) is very heavy ($\approx$ 18.6 g/cm²) and very thick ($\approx$ 12.9 cm). Furthermore, predicted surface recession is very large ($\approx$ 9.6 cm). It is doubtful that tape-wrapped CP could be fabricated to these thicknesses with adequate quality. Such thicknesses were achieved for the nosecap on the Galileo probe, but that was a chopped molded construction of a relatively small nosecap. To fabricate a one-piece chopped molded heatshield for this large Neptune aerocapture vehicle would require presses of a size that do not currently exist. The option is a buildup of chopped molded tiles, but using tiles and the requisite gap fillers at such severe conditions introduces additional risk. In comparison, the Genesis concept has marginally lower areal weight ($\approx$ 17.9 g/cm^2), but is even thicker ($\approx$ 14.9 cm). The sizing analysis determined that the C-C facesheet needed to be $\approx$ 8.9 cm thick backed by $\approx$ 6 cm of carbon fiberform. Fabricating a cloth-reinforced carbon-carbon to that thickness with good uniformity and quality is a challenge. The predicted surface recession is also large ($\approx$ 7.6 cm), but less than CP. The TPS requirements at "Low" condition heating are about 2/3rds those at the reference condition. It should be noted that for the evaluation of the "Low" condition, a reduced density carbon phenolic (ρ = 0.96 g/cm^3) was included in the evaluation. Such materials were fabricated and evaluated for other TPS applications in the 70s and 80s, but never used. The reduced density CP looks relatively competitive at the "Low" condition, but it is doubtful it could withstand peak heat fluxes in the range of 10 kW/cm^2 without spalling. Given these results, it was decided to select fully dense carbon phenolic as the reference TPS in the nose region.

WINDSIDE CENTERLINE (ZONE 2)

The "Low" and "Med" aeroheating rates and loads along Monte Carlo trajectory #1647 are shown in Figure 9. As was the case with the nose region, after further aeroheating analyses, it was concluded that the "High" values were outside of the expected range and the "Med" level of aeroheating was used for TPS sizing for the reference vehicle. The peak heat fluxes are in the range 3-6 kW/cm^2, eliminating lower density candidate materials. The total heat loads are still very large – 300-700 kJ/cm^2. TPS sizing in this region was done for fully dense carbon phenolic, the Genesis concept and reduced density carbon phenolic. As seen in Figure 10, the reduced density carbon phenolic has the lowest areal weight ($\approx$ 6.5 g/cm^2 at the reference condition), but is very thick ($\approx$ 6.2 cm). The surface recession is the largest among the materials considered ($\approx$ 3.9 cm). The Genesis concept's areal weight is marginally greater (6.9 g/cm^2) and the thickness is very large ($\approx$ 8.7 cm). The sizing analysis determined that the C-C facesheet needed to be $\approx$ 3.1 cm thick backed by $\approx$ 5.6 cm of carbon fiberform. However, the surface recession is the smallest among the materials considered ($\approx$ 2.0 cm). In comparison, fully dense carbon phenolic is the heaviest solution (8.9 g/cm^2 at the reference condition). Moreover, the TPS requirements at the low condition are more than half those at the reference condition. Given these results, the reduced density carbon phenolic was selected as the reference TPS in Zone 2. The major uncertainty is whether reduced density carbon phenolic can handle these peak heat fluxes reliably.

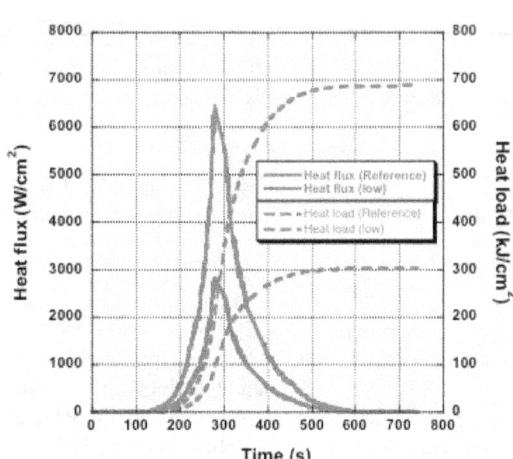

Figure 9. Total low and medium (Reference) heat rates and heat loads for windside centerline region (Zone 2).

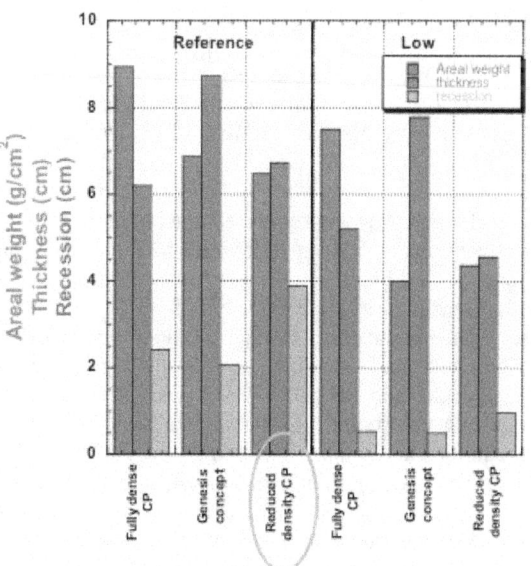

Figure 10. TPS sizing along the windside centerline Zone 2).

127

LEE SIDE (ZONES 3 AND 4)

For the lee side of the vehicle, the "Low", "Med", and "High" aeroheating rates and loads along Monte Carlo trajectory #1647 are shown in Figure 11. The "High" values are almost the same as the "Med" values and were eliminated from further consideration. The "Med" level of aeroheating was used for TPS sizing in Zone 3, just aft of the nose on the lee side. Peak heating rates in that region are ≈ 625 W/cm^2 and the total heat load is ≈ 68 kJ/cm^2.

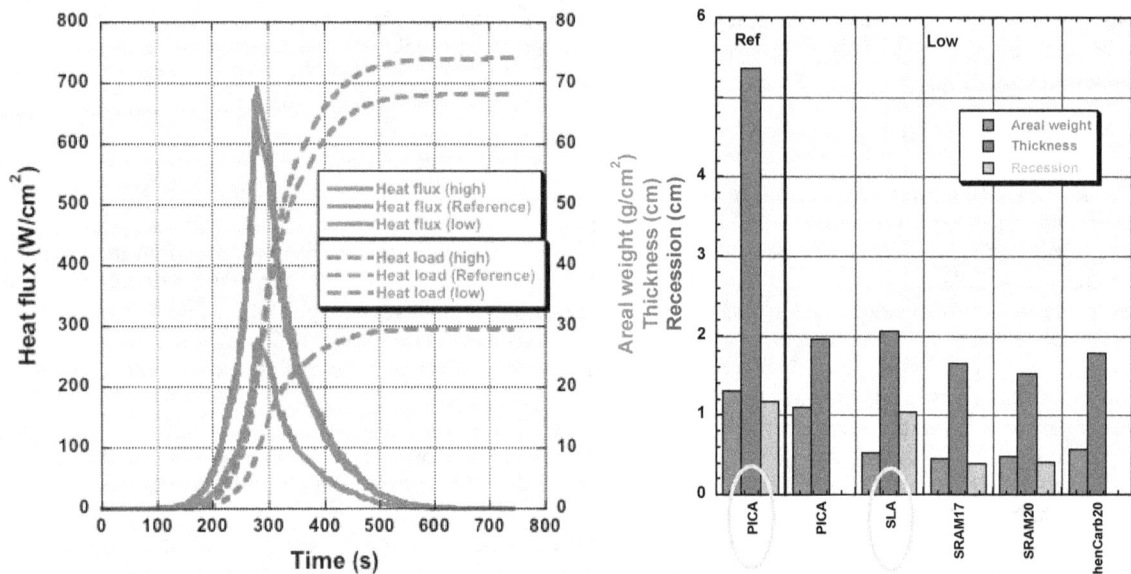

Figure 11. Total heat rates and heat loads on the lee side (Zones 3 and 4).

Figure 12. TPS sizing on the lee side (Zones 3 and 4).

Unfortunately, there are not many low-density TPS materials that can handle heat fluxes of that magnitude efficiently. The lack of low-mid density materials with such a capability is a serious deficiency in the TPS arsenal. Consequently, the only candidate TPS material evaluated for Zone 3 applications was PICA (Phenolic Impregnated Carbon Ablator), with a density of ≈ 0.240 g/cm^3.

The PICA sizing results, shown in Figure 12, illustrate that the areal weight is moderate (≈ 1.29 g/cm^2). It is still rather thick (≈ 5.36 cm) but within manufacturing capabilities. Surface recession is predicted to be comparatively small (≈ 1.17 cm).

The low heat flux heating solutions were applied to Zone 4, where the peak heat flux is only 272 W/cm^2 and the total heat load is ≈ 30 kJ/cm^2. In this region TPS sizing calculations were done for PICA, SLA-561V, SRAM17, SRAM20, and PhenCarb20. As seen, PICA would be a heavy and thick TPS solution for this relatively modest heating environment, although no surface recession is predicted. The difference in areal weight and thickness requirements among SLA-561V, SRAM17 and SRAM20 is not negligible, but still relatively small. Predicted surface recession is greatest for SLA-561V. In comparison, no recession is predicted for PhenCarb20 and its areal weight and thickness requirements are similar to the silicone-based low-density ablators. The selection of SLA-561V for the reference design was due solely to its flight heritage.

SUMMARY

This study of a Neptune aerocapture mission has demonstrated that for the geometry and mission design parameters selected, it is a very challenging mission from the TPS standpoint. The windside heating rates are so high (particularly in the stagnation region) that only fully dense carbonaceous TPS materials can handle them with reliable performance. Furthermore, the windside heat loads are enormous. These long duration aerocapture trajectories require extremely thick TPS, particularly on the wind side. Such carbonaceous materials are not good insulators, but are good ablators, and the results demonstrate that, on the windside, ablation is the dominant mechanism. Since carbon is the best ablator we currently have, whether improvement could be attained with development of new TPS materials is debatable. However, the required thickness of a tape-wrapped carbon phenolic and/or a cloth-reinforced carbon-carbon cannot be manufactured as a high quality product. It may be possible to manufacture as a stackup of 2.5-5.0 cm thick sheets, mechanically joined. However, that would require some manufacturing development. It is also important to appreciate that the effects of ablation and shape change on vehicle aerodynamics were not considered and should be investigated, as they may be very important.

ACKNOWLEDGMENTS

The authors wish to express their appreciation to William M. Congdon of Applied Research Associated who did the TPS sizing analyses for SLA-561V, SRAM17, SRAM20 and PhenCarb20 on the lee side (Zone 4). The SRAM and PhenCarb materials are proprietary products of ARA and, consequently, ARA has the only analysis models for these materials. In addition, ARA has an analysis model for SLA-561V (developed from arc jet test data obtained during the Mars Pathfinder project) that addresses surface recession, whereas the current ARC model does not. Interestingly, the thickness requirements for SLA defined by ARA (with consideration of recession) and those defined by ARC (without consideration of surface recession) were essentially identical.

REFERENCES

[1] Lockwood, M.K., "Neptune Aerocapture Mission Systems Analysis," AIAA-2004-4951, AIAA Atmospheric Flight Mechanics Conference, Providence, RI, August 16-19, 2004.

[2] Noca, M., "Mission Analysis," presented at In-Space Propulsion Aerocapture Project's Neptune Systems Analysis Review, Huntsville, AL, 28 October 2003.

[3] Justus, C., Duvall, A. and Keller, V., "Atmospheric Models for Aerocapture Systems Studies," AIAA-2004-4952, AIAA Atmospheric Flight Mechanics Conference, Providence, RI, August 16-19, 2004.

[4] Edquist, K., Prabhu, R., Hoffman, D. and Rea, J. "Configuration, Aerodynamics and Stability Analysis for a Neptune Aerocapture Orbiter," AIAA-2004-4953, AIAA Atmospheric Flight Mechanics Conference, Providence, RI, August 16-19, 2004.

[5] Gnoffo, P.A., "An Upwind Based Point Implicit Relaxation Algorithm for Viscous Compressible Perfect Gas Flows," NASA TP 2953, Feb. 1990.

[6] Cheatwood, F.M. and Thompson, R.A., "The Addition of Algebraic Turbulence Modeling to Program LAURA," NASA TM 107758, Apr. 1993.

[7] Cheatwood, F.M. and Gnoffo, P.A., "User's Manual for the Langley Aerothermodynamic Upwind Relaxation Algorithm (LAURA)," NASA TM-4674, Apr. 1996.

[8] Starr, B., Powell, R., Westhelle, C. and Masciarelli, J., "Aerocapture Performance Analysis for a Neptune- Triton Exploration Mission," AIAA-2004-4955, AIAA Atmospheric Flight Mechanics Conference, Providence, RI, August 16-19, 2004.

[9] Wright, M.J., Candler, G.V. and Bose, D., "Data-Parallel Line Relaxation Method for the Navier-Stokes Equations," AIAA Journal, Vol. 36, No. 9, pp. 1603-1609, Sep. 1998.

[10] Whiting, E.E., Park, C., Liu, Y., Arnold, J.O. and Paterson, J.A., "NEQAIR96, Nonequilibrium and Equilibrium Radiative Transport and Spectra Program: User's Manual," NASA RP-1389, Dec. 1996.

[11] Hollis, B., Olejniczak, J., Wright, M., Takashima, M., Sutton, K., and Prabhu, D., "Preliminary Convective- Radiative Heating Environments for a Neptune Aerocapture Mission," AIAA-2004-5177, AIAA Atmospheric Flight Mechanics Conference, Providence, RI, August 16-19, 2004.

[12] Nicolet, W.E., "Advanced Methods for Calculating Radiation Transport in Ablation-Product Contaminated Boundary Layers," Report No. 69-61, Aerotherm Corporation, Mountain View, California, December 27, 1969.

[13] Nicolet, W.E., "User's Manual for the Generalized Radiation Transfer Code (RAD/EQUIL)," Report No. UM-69-9, Aerotherm Corporation, Mountain View, California October 1, 1969.

[14] Chen, Y. -K. and Milos, F.S., "Ablation and Thermal Analysis Program for Spacecraft Heatshield Analysis", *Journal of Spacecraft and Rockets*, vol. 36, No. 3, 1999, pp. 475-483.

STRUCTURAL DESIGN FOR A NEPTUNE AEROCAPTURE MISSION

R. Eric Dyke
Swales Aerospace, NASA Langley Research Center, Hampton, VA 23681

Glenn A. Hrinda
NASA Langley Research Center, Hampton, VA 23681

A multi-center study was conducted in 2003 to assess the feasibility of and technology requirements for using aerocapture to insert a scientific platform into orbit around Neptune. The aerocapture technique offers a potential method of greatly reducing orbiter mass and thus total spacecraft launch mass by minimizing the required propulsion system mass. This study involved the collaborative efforts of personnel from Langley Research Center (LaRC), Johnson Space Flight Center (JSFC), Marshall Space Flight Center (MSFC), Ames Research Center (ARC), and the Jet Propulsion Laboratory (JPL). One aspect of this effort was the structural design of the full spacecraft configuration, including the ellipsled aerocapture orbiter and the in-space solar electric propulsion (SEP) module/cruise stage. This paper will discuss the functional and structural requirements for each of these components, some of the design trades leading to the final configuration, the loading environments, and the analysis methods used to ensure structural integrity. It will also highlight the design and structural challenges faced while trying to integrate all the mission requirements. Component sizes, materials, construction methods and analytical results, including masses and natural frequencies, will be presented, showing the feasibility of the resulting design for use in a Neptune aerocapture mission. Lastly, results of a post-study structural mass optimization effort on the ellipsled will be discussed, showing potential mass savings and their influence on structural strength and stiffness

NOMENCLATURE

Al	=	aluminum	*LaRC*	=	Langley Research Center
ARC	=	Ames Research Center	*L/D*	=	lift to drag ratio
AU	=	astronomical units	*MAC*	=	mass acceleration curve
B/S	=	backshell	*MEL*	=	master equipment list
CBE	=	current best estimate	*MS*	=	margin of safety
CG	=	center of gravity	*MSFC*	=	Marshall Space Flight Center
F/B	=	forebody	*NSM*	=	non-structural mass
FEA	=	finite element analysis	*OML*	=	outer mold line
FEM	=	finite element model	*PAF*	=	payload adapter fitting
FS	=	factor of safety	*PM*	=	propulsion module
Gr	=	graphite	*SA*	=	solar array
HGA	=	high gain antenna	*SEP*	=	solar electric propulsion
JPL	=	Jet Propulsion Laboratory	*TPS*	=	thermal protection system
JSFC	=	Johnson Space Flight Center	*Xe*	=	Xenon

INTRODUCTION

Structural sizing for a conceptual aerocapture spacecraft to Neptune was required to establish concept feasibility and to obtain preliminary component mass estimates. The full spacecraft launch stackup consisted of an ellipsled aerocapture/orbiter vehicle sitting atop a propulsion module (PM)/cruise stage, all designed to fit within the 5 meter fairing of a Delta IV Heavy launch vehicle[1]. The PM/cruise stage contained the solar arrays (SA's), Xenon (Xe) tank and other subsystems for the 30 kW, 6-engine solar electric propulsion (SEP) system to be used out to 3 AU. It also held a small hydrazine fuel tank, telecommunication antennae, navigation equipment, thermal radiators, and two Neptune direct entry atmospheric probes which were considered simple lumped masses for this study.

There were four basic objectives for the structural analysis: 1) Support all science payload and subsystem components in the required volume, 2) Meet minimum stackup natural frequencies at launch, 3) Sustain structural stresses at launch and during aerocapture with acceptable margins of safety (MS), and 4) meet the above three objectives with minimal structural mass. Objective 1) above was accomplished by multiple packaging/analysis iterations between JPL and LaRC personnel, producing several ellipsled orbiter configurations and overall size changes before an acceptable design was found. Launch loading criteria from the Boeing Payload Planners Guide[2] and aerocapture loading criteria from NASA Langley Monte Carlo simulations[3] were used in conjunction with the commercially available finite element analysis (FEA) software I-DEAS[4] to size structure with acceptable strength and stiffness to meet objectives 2) and 3) above. I-DEAS FEA and hand calculations were used to size the ellipsled orbiter and the PM/cruise stage during the scheduled design/analysis cycle. Due to challenges in packaging all of the required payload instruments and subsystem components to meet design functionality and overall center of gravity (CG) requirements, and to the ensuing shortened time available for analysis, mass optimization was performed after the scheduled design/analysis cycle. The commercially available software HyperSizer™ [5] was used to help reduce mass on the ellipsled orbiter. No similar mass optimization effort was done on the PM/cruise stage.

The resulting structure consists of a composite material honeycomb sandwich construction ellipsled orbiter aeroshell surrounding a deep-rib stiffened honeycomb sandwich payload deck. The ellipsled orbiter aeroshell is separate forebody (F/B) and backshell (B/S) pieces integrally stiffened with longitudinal and circumferential blades. The F/B and B/S separate from the payload deck after aerocapture via several pyrotechnic separation fittings. The resulting PM/cruise stage is a stiffened Al skin with Al rings and trusses to support the hydrazine and Xe tanks and the two direct entry probes and an Al frame to support the SEP engines.

FUNCTIONAL REQUIREMENTS

ORBITER SHAPE SELECTION

Neptune atmosphere profiles developed by Justus, Duvall, and Keller[6] at MSFC and Neptune atmosphere entry parameters developed by JPL[7] and LaRC[3] personnel were used to determine the required aerocapture vehicle shape and aerodynamic characteristics to meet the stringent entry corridor needed for aerocapture at Neptune[3]. Edquist[8] (LaRC) evaluated the aerodynamics of several entry vehicle shape classes, including sphere-cone, biconic, bent biconic, and ellipsled, to find an appropriate shape giving the necessary volume and aerodynamic lift to drag ratio (L/D). The resulting vehicle, as shown in Fig. 1, was an ellipsled shape with a flattened bottom. The general ellipsled shape is a body of revolution with an ellipsoid nose and circular cylinder aft end. The flattened ellipsled has an upper portion that is half a body of revolution and a lower portion that is a general ellipsoid nose and elliptical aft cylinder.

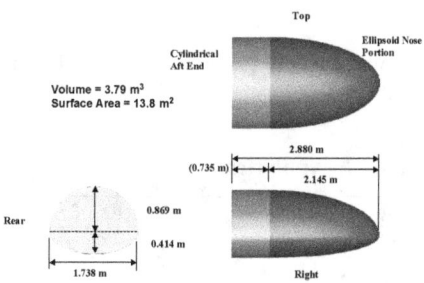

Figure 1. Flattened Ellipsled Geometry

ORBITER AND PM/CRUISE STAGE REQUIREMENTS

The primary functions of the ellipsled orbiter were to provide the aerodynamic shape necessary to facilitate orbiter aerocapture using the Neptunian atmosphere, to provide the volume necessary to package the scientific payload and other subsystems, to provide sufficient structural MS for natural frequency, buckling, and static stress for launch and aerocapture loading, and to do all of the above with minimal structural mass and complexity. There were several challenges to overcome in fulfilling these functions. The ellipsled aeroshell structure had to support a high thermal protection system (TPS) mass[9] due to the high aeroheating during aerocapture[10]. The numerous payload and other subsystem components had to be packaged to allow their proper functions but also to provide proper overall mass CG to maintain the required ellipsled angle of attack for aerodynamic control and stability during the full aerocapture phase[8, 11]. There were also large variations in aerocapture g loads during the course of the conceptual design phase. These challenges required multiple ellipsled sizing iterations as detailed in a later section.

The primary functions of the PM/cruise stage were to support the ellipsled during launch and cruise; to provide attachments for the two direct entry Neptune probes, telecom antennae, thermal control radiators, and SEP system components; to provide sufficient structural MS for natural frequency, buckling, and static stress for launch loading; and to do all of the above with minimal structural mass and complexity. The primary challenge for the PM/cruise stage structure was providing for the numerous component attachments in a compact design without compromising their proper functions. The two direct entry probes required specific alignment to allow separation independent of each other and the ellipsled and to allow separation along a vector going through (or as close as possible to) the vehicle CG. The large ellipsled mass sitting on top during launch also required extra PM/cruise stage stiffness to meet the stackup launch natural frequency requirements.

STRUCTURAL ANALYSIS REQUIREMENTS

DESIGN LOADS

Design launch loads were taken from the Boeing Payload Planners Guide for the Delta IV Heavy[2], and are summarized in Table 1 below. For the static analysis and natural frequency calculations, the full stackup was assumed restrained at the payload adapter fitting (PAF). Aerocapture design g loads were taken from the 3-sigma g loads from the Monte Carlo entry analysis[3], and were balanced with aeropressure loads on the ellipsled aeroshell using an unpublished coarse pressure distribution from N. Takashima (AMA/LaRC) dated September 12, 2003. Component level loads from mass acceleration curves (MAC's), and sine, random, and acoustic loading were not analyzed as part of this study.

Table 1. Static Load Factors

Event	Loading
Launch	6.0 g's axial + 0.5g's lateral, any direction
	2.3 g's axial + 2.0 g's lateral, any direction
Aerocapture	22.1 g's, acting 11.3 degrees aft of vertical relative to ellipsled payload deck

STRENGTH AND STIFFNESS

Standard strength and stability factors of safety (with verification) listed in Table 2 below were used in the structural analysis.

Table 2. Analytical Factors of Safety

Mode	Factor of Safety
Metallic ultimate stress	1.4
Metallic yield stress	1.25
Stress in composites	1.4
Buckling	1.5

Stackup minimum required natural frequencies at launch, taken from the Boeing Payload Planners Guide for the Delta IV Heavy[2], were >8 Hz for the fundamental lateral modes, and >30 Hz for the fundamental axial mode.

ANALYSIS METHODS

Standard "stick and panel" finite element model (FEM) construction with 2-D (non-solid) elements was used for all structural analyses. Components such as the two direct entry probes, radiators, science instruments, fuel tanks/fuel, etc., were modeled as lumped masses and connected to the vehicle structure using rigid-type element connectors or beam elements as appropriate. All FEM's were constructed with I-DEAS, and solved with I-DEAS (2.88m ellipsled) or NASTRAN (5.5m ellipsled) as described below.

The structural analysis was done in two phases. First, the ellipsled was analyzed using the aerocapture pressure loads with an inertia relief solution method that balances the pressures with entry g loads. The TPS was modeled as non-structural mass (NSM) on the aeroshell elements using areal densities provided by B. Laub (ARC)[9] with 30% growth factors applied. For the structural analysis, the F/B TPS areal density (55.4 kg/m^2) and B/S and base TPS areal density (5.54 kg/m^2) were each assumed constant, making two TPS zones. Later TPS analysis modified this to four TPS zones[12] in an effort to help reduce TPS mass, but was not available in time for this structural analysis. Instruments and other subsystem components were modeled as lumped masses with 30% growth factors applied. Non-point masses such as thermal blankets, cabling, etc., were added to the payload deck as NSM with 30% growth factors applied. The ellipsled aeroshell and payload deck structure were then sized and the resulting structure masses were considered current best estimate (CBE).

For the full stackup at launch, the ellipsled structure mass was adjusted to include the 30% growth factor, with the growth portion being applied as NSM to the existing structure plate elements. The SEP/cruise stage payload components (radiators, probes, fuel tanks, etc.) were modeled as lumped masses with the 30% growth factors applied. Non-point masses such as cabling, etc., were added as NSM to the cruise stage cylinder and thrust tube. The stackup structure was then sized for the launch loads, and the resulting structure masses for the PM/cruise stage were considered CBE. After the preliminary structure sizing for static loads, the ellipsled was evaluated for buckling under aerocapture loads. The full stackup was evaluated for natural frequency and buckling in the launch configuration under launch loads.

ORBITER SIZE ITERATIONS

5.5 M ELLIPSLED DESIGN

The ellipsled aeroshell was initially 5.5m long, maximized to fit in a Delta IV Heavy 5m fairing[2]. The length was determined by ratioing the maximum aeroshell width that could fit inside the Delta IV fairing. This provided the largest orbiter volume for science payloads and greatest width for mounting a rigid high gain antenna (HGA). Different internal structures to support the rigid aeroshell and mount payloads were tried. Figure 2 shows an early concept using a space truss to maintain the outer mold line (OML) of the aeroshell.

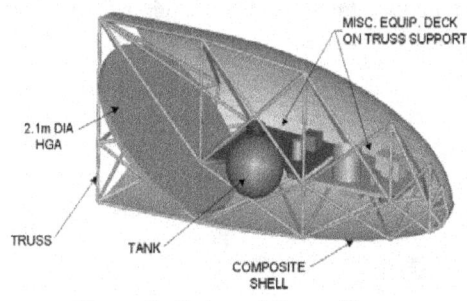

Figure 2. Internal Space Truss

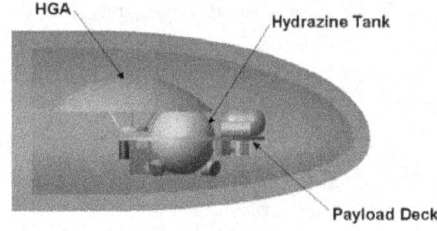

Figure 3. Internal Payload Deck

This configuration relied on the trusses for all equipment mounting and did not require large stiffening of the shell. The load path from all payload and aeroshell mass continued through the space truss into a cruise stage adapter. The purpose of using the space truss was to minimize aeroshell mass with an efficient, highly stiff internal support system. As the design study proceeded, the payload requirements and their configurations inside the ellipsled were constantly being revised. As a result, the internal truss design became difficult to alter while trying to package the rigid HGA within the trusses. A second method was tried that used a flat, stiffened deck for mounting the payload. The flat payload platform offered a convenient surface for securing equipment and also allowed for quick component configuration changes. Figure 3 shows the flat payload deck and major components of the orbiter.

A single hydrazine tank was located near the ellipsled CG with a rigid HGA mounted as shown. The rigid antenna was oriented to fit inside the aeroshell and mounted to the payload deck. The load path for this concept had the aeroshell supporting the payload deck during launch. All loads would then be taken into an elliptic thrust adaptor and continue through to the cruise stage. The cruise stage configuration during this time of the design study was unknown so a cruise stage from an earlier design study[13] was used. Figure 4 shows the FEM of the ellipsled with its adapter and cruise stage.

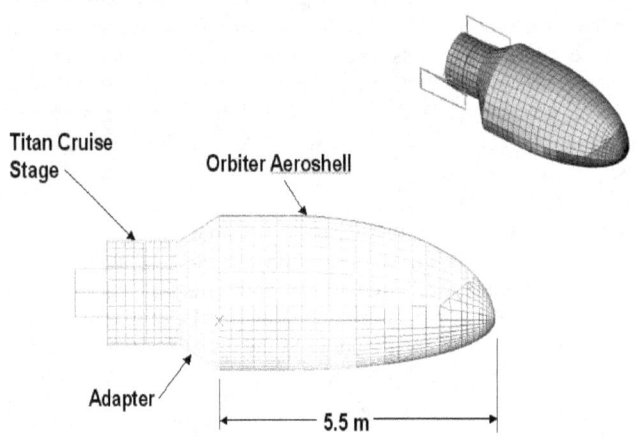

Figure 4. 5.5m Ellipsled with Preliminary Adapter/Cruise Stage

5.5M ELLIPSLED STRUCTURAL ANALYSIS

The 5.5m ellipsled aeroshell structure was analyzed using standard FEA combined with a non-deterministic structural sizing program called HyperSizer™ which allows many trial composite sections and materials to be analyzed very efficiently using only one coarsely meshed FEM. The HyperSizer™ analysis started with a coarse NASTRAN[14] FEM of the full stack shown in Fig. 4, subjected to launch loads. That FEM, containing only CQUAD4, CTRIA3, CONM2, and CBAR NASTRAN elements, was solved with NASTRAN and the mesh and resulting element internal loads were imported to HyperSizer™. Figures 5 and 6 show how the FEM was divided into major components reflecting the mission of the orbiter. HyperSizer™ did not require structure remeshing to reflect structural changes necessary to support changing payload components from the master equipment list (MEL).

Detailed finite element modeling of panel stiffening methods was not necessary. Within HyperSizer™, a user

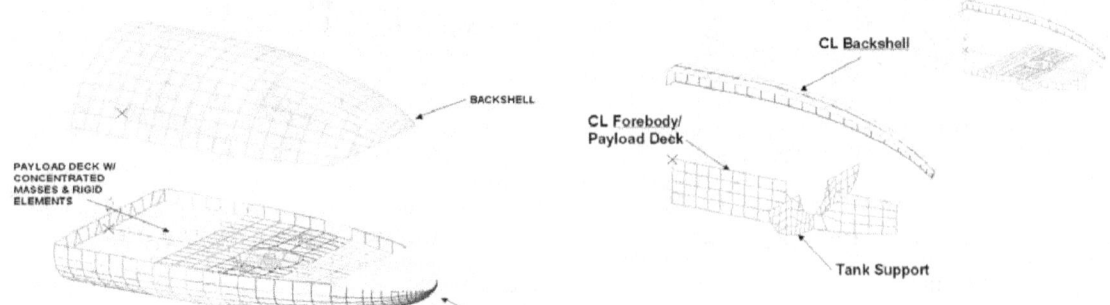

Figure 5. Major Aeroshell Components Figure

Figure 6. Internal Stiffening Structure

can choose among many common aerospace structural concepts such as blade-stiffened panels, honeycomb core panels and isogrids while still using the same coarse FEM. Figure 7 shows the analysis path taken by HyperSizer™. Each color shown in the figure represents a group of finite elements with common NASTRAN property and material cards, lumped together as a component (or "panel") when imported into HyperSizer™. The figure shows the payload deck divided into four components that will each be sized for optimal panel stiffening method, thickness, and material.

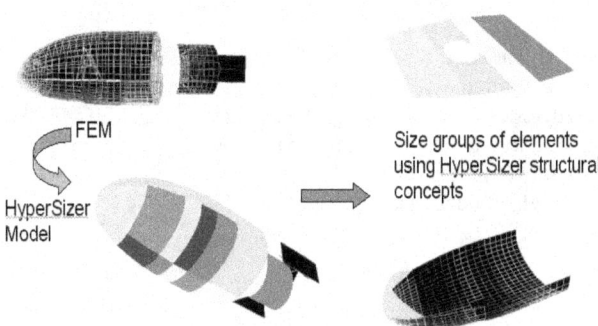

Figure 7. HyperSizer™ Analysis Path

A F/B and B/S were created and attached together at locations where they would separate after aerocapture. Groups of finite elements were created for optimizing in HyperSizer™ and are shown as different colors in Figs. 4-6. Stiffening of the payload deck and aeroshell became necessary as the analysis proceeded. Figure 6 shows a major bulkhead required to support the Hydrazine tank and axial stiffeners to help transfer loads during launch. TPS mass was input into HyperSizer™ as NSM and could easily be altered to suit different thermal material trade studies. Launch and aerocapture loading and structure stiffness requirements from Table 1 in the Design Loads section were used. An older cruise stage FEM from a previous design study[13] was used to obtain estimated full stack stiffness to check launch-configuration natural frequencies. Subsystem/payload component masses from the latest MEL were lumped on the stiffened payload deck based on the latest design. Components were constantly being moved during the design, causing many modifications to the analysis. Later in the study the HGA antenna was replaced with a deployable antenna. This decision drastically affected the aeroshell design since the maximum geometry to fit the rigid HGA was no longer required. The aeroshell volume could be shrunk to minimize structural mass. This led to a final design concept requiring a 2.88m long ellipsled that also represented the most current MEL. The design study was then divided into two paths: one using a 5.5m long ellipsled and the other using a 2.88m ellipsled that also represented the most current design and MEL. The purpose for having two design concepts was to provide maximum and minimum structural mass estimates for the systems study. The 5.5m ellipsled design was finished to give mission planners a maximum structural mass and internal volume if a larger ellipsled is required.

5.5M ELLIPSLED STRUCTURAL ANALYSIS RESULTS

The final structural member masses for the 5.5m ellipsled are shown in Table 3 below[15], followed by more detailed construction descriptions. These results were considered worst case structural mass estimates for the given aerocapture mission to Neptune.

Table 3. 5.5m Ellipsled Component Masses

Component	Area (m²)	Structural Mass (kg)	TPS Mass (kg)
Heatshield (F/B)	22.47	210.54	1245.35
B/S	21.30	151.57	118.02
Payload Deck	12.88	271.48	No TPS
Aft Bulkhead	6.37	40.36	35.28
Totals		673.95	1398.65

Heatshield (F/B)– 5.08 cm thick with a Hexcell 5052 Alloy Hexagonal Al Honeycomb core and 1.651 mm Gr-Polyimide face sheets

B/S– 3.39 cm thick with a Hexcell 5052 Alloy Hexagonal Al Honeycomb core and 1.651 mm Gr-Polyimide face sheets

Aft Bulkhead – 2.54 cm thick with a Hexcell 5052 Alloy Hexagonal Al Honeycomb core and 1.651 mm Gr Polyimide face sheets

Payload Deck –Al Isogrid The two lowest lateral stack modes were 17.51 Hz and 17.93 Hz. The lowest axial mode was 49.98 Hz, involving structure for the orbiter thrusters. All local buckling checks were performed within HyperSizer™.

Honeycomb core with facesheets was used for the overall aeroshell design. It provided the lowest mass that met all stress and dynamic modes criteria for the aeroshell. The isogird design shown in Fig. 8 was selected for the scientific payload platform. The detailed geometry would have been difficult and time consuming to create with a typical FEA. HyperSizer™ was able to quickly show a payload deck isogrid design that is well suited for mounting components with ample openings for running cables and piping. The isogird design mass was roughly the same as that required for a blade stiffened payload deck using honeycomb.

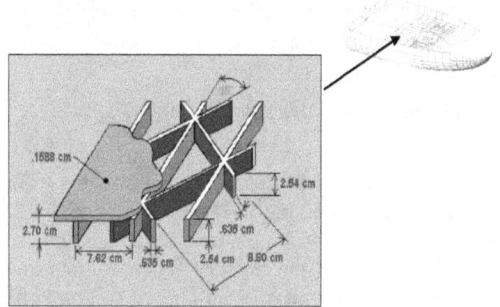

Figure 8. Payload Deck Isogrid Design

As mentioned above, preliminary mass estimates and HGA design changes allowed the ellipsled to be reduced to 2.88m. Figure 9 shows a size/design comparison between the original, larger 5.5m aeroshell with old cruise stage, and the revised, smaller 2.88m ellipsled with new cruise stage, described more fully in the next sections.

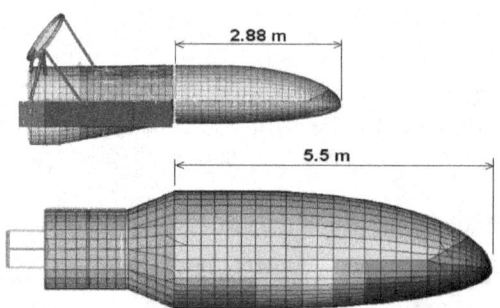

Figure 9. 5.5m and 2.88m Ellipsled Comparison

2.88M ELLIPSLED DESIGN

After initial structure and TPS mass estimates showed unacceptably high values for the 5.5m ellipsled, and changes were made to use a deployable HGA, a parallel analysis effort was started to size a smaller ellipsled. The ellipsled was reduced to 3.5 m, then 3.2m, then finally 2.88m. Figures 10[16] and 11 show the full stackup design and its FEM, respectively, for the 2.88m ellipsled in the Delta IV Heavy 5m fairing. Figures 12[16] and 13 show the ellipsled orbiter design, with major functional components, and its FEM, respectively.

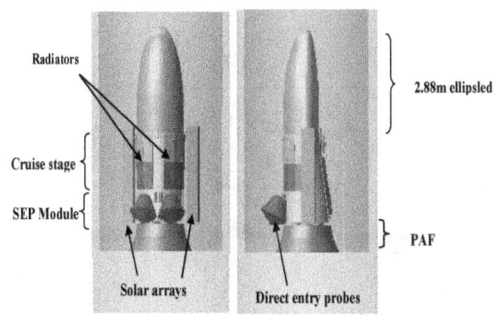

Figure 10. Full Stackup with 2.88m Ellipsled in 5m Delta IV Heavy Fairing

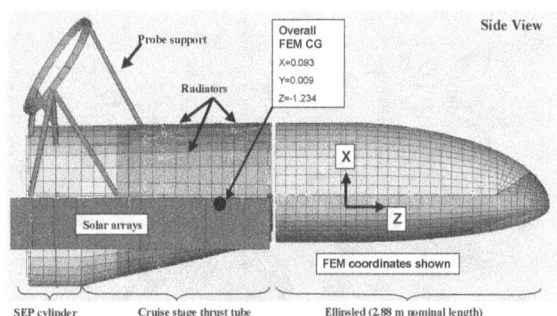

Figure 11. Full Stackup FEM

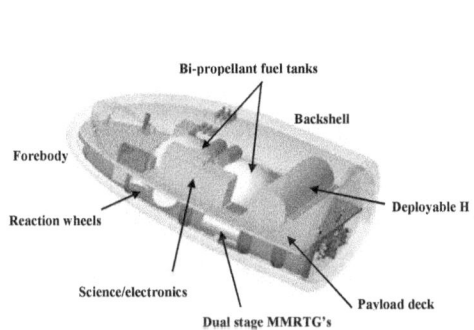

Figure 12. 2.88m Ellipsled Orbiter

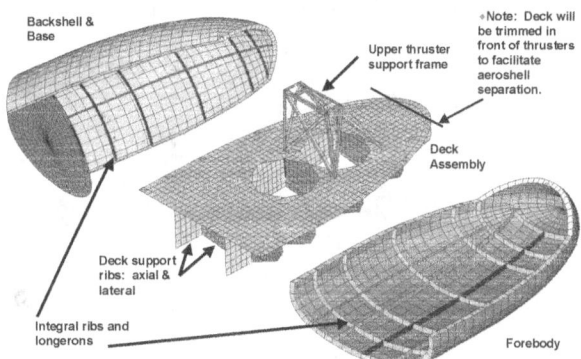

Figure 13. 2.88m Ellipsled Orbiter FEM

The F/B and B/S base are uniform 2.54 cm thick sandwich structure with 5052 Al honeycomb and 0.132 cm Gr-Polyimide facesheets, stiffened with 0.318 cm thick integral monolithic composite blade longerons and circumferential ribs. The payload deck is also a 2.54 cm thick sandwich structure with 5052 Al honeycomb and 0.132 cm Gr-Polyimide facesheets. It is stiffened with full-depth lateral and longitudinal sandwich structure ribs, 1.27 cm thick with 5052 Al honeycomb and 0.132 cm Gr-Polyimide facesheets. The bi-propellant fuel tanks are further supported by small Al tube struts under the deck. The upper frame is 2.54 cm x 2.54 cm x 0.130 cm Al angles, and supports thrusters for on-orbit attitude control. The ellipsled is tied to the PM/cruise stage with eight pyrotechnic fittings which separate the ellipsled from the cruise stage prior to aerocapture. The payload deck is tied to the F/B and B/S base with twenty separation fittings which fire after aerocapture to separate the F/B and B/S, leaving the payload deck on orbit. During aerocapture, the component inertia loads from the orbiter's high-g deceleration are transmitted across the payload deck panels, into the ribs, then into the aeroshell (primarily the F/B), where they are balanced by the aeropressure loads on the aeroshell exterior.

Figures 14[16] and 15 show the PM/cruise stage design with functional components, and its FEM representation with major structural components, respectively. Both the SEP cylinder and cruise stage thrust tube are stiffened skin construction. The 0.254 cm Al skin is stiffened by a series of Al longerons and rings, as shown in Fig. 16, which transmit launch loads into the PAF and provide hard points for component attachments such as the hydrazine and Xe tanks, SA's, radiators, etc. An Al ring frame at the bottom of the SEP cylinder, stiffened by 5.08 cm Al tube struts, provides attach points for the six SEP engines. The two entry probes are supported by 5.08 cm Al channel-section rings with 5.08 cm Al tube trusses. The Xe tank is supported by a 5.08 cm Al channel-section ring and 5.08 cm Al tube struts at the bottom, and 2.54 cm Al tube struts at the top. The hydrazine tank is supported by a single Al ring with stiffening struts. During launch, the ellipsled inertia loads enter the PM/cruise stage via the eight separation fittings. The inertia loads from the individual PM/cruise stage components enter the stiffened skin structure through their respective support structure. All of these loads are then transmitted down the stiffened skin, eventually being reacted at the PAF.

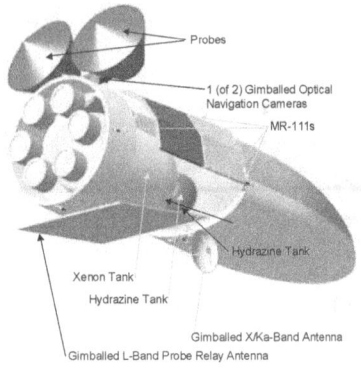

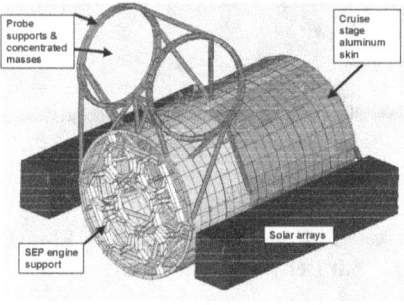

Figure 14.   PM/Cruise Stage Components   **Figure 15.   PM/Cruise Stage FEM**

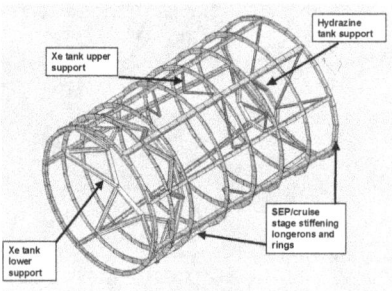

**Figure 16.   PM/Cruise Stage FEM
showing internal longerons and rings**

2.88M ELLIPSLED STRUCTURAL ANALYSIS RESULTS

The FEM modal analysis showed that both the ellipsled and PM/cruise stage structures were largely stiffness critical, and were thus primarily sized to maintain the minimum design natural frequencies during launch. The full depth ribs on the ellipsled payload deck and their attachment to the aeroshell F/B kept the local deck natural frequency above 32 Hz. The minimum natural frequencies for the full stackup at launch were 11.08 Hz lateral and 32.01 Hz axial, above the 8 and 30 Hz requirements[2], respectively. The lateral mode involved the full stack bending in the "weakwise" direction, normal to the payload deck, while the axial mode involved the SEP engines and their support structure "bouncing" in the direction of the stackup longitudinal axis.

Since the structure was largely stiffness critical, the FEM static analyses showed generally high structural margins of safety, with only a few local high stress areas. On the ellipsled, the areas of lowest MS were the F/B longeron strength at the B/S separation fitting interface during aerocapture, and the propellant tank support strut buckling at aerocapture. On the PM/cruise stage, the lowest MS was against bending of the SA base support during launch. The maximum static deflection of 0.66 cm occurred at the ellipsled nose during launch for the maximum lateral g condition. An I-DEAS eigenvalue buckling solution of the full stackup showed a buckling margin of safety of 2.47, with the critical location being the upper Al skin panel on the cruise stage.

Summaries of the ellipsled alone and full stackup masses are shown in Tables 4 and 5 below. Table 4 shows the ellipsled alone evaluated for aerocapture loading. The total mass of 1412 kg includes 474.2 kg of TPS mass and 136.5 kg of CBE structure mass.

Table 4. Ellipsled Only Mass Summary: Aerocapture Evalutaion

Item	Mass (kg)
Forebody	464.4
F/B TPS (growth)	419.5
F/B structure	44.9
Backshell	84.9
B/S TPS (growth)	42.1
B/S structure	42.8
Base	19.75
Base TPS (growth)	12.6
Base structure	7.15
Deck	191.6
Deck NSM (growth)	170
Deck structure	21.6
Deck ribs structure	17.9
Lumped masses (growth)	631.2
Tanks, etc.	606.5
Separation fittings	24.7
Tank support rods	0.42
Thruster support frame	1.75
Total TPS Mass (growth)	**474.2**
Total Structure Mass (CBE)	**136.5**
Total Ellipsled Mass	**1412**

In Table 5, the total stackup mass of 4190 kg includes 1460 kg for the ellipsled (which includes the 30% growth factor applied to the CBE structure mass from above) and 2730 kg for the PM/cruise stage. The PM/cruise stage mass includes 203.82 kg of CBE structure mass. For the full system analysis mass tracking, the CBE values are increased by 30% for growth values, giving a total structure mass for the stackup at launch of 442 kg.

Table 5. Full Stackup Mass Summary at Launch

Components				Mass (kg)
Ellipsled Total				**1460**
Aeroshell total				597.81
	Forebody total			478.1
		F/B TPS		*419.5*
		F/B structure		*58.6*
	Backshell total			97.81
		B/S TPS		*42.12*
		B/S structure		*55.69*
	Base total			21.9
		Base TPS		*12.61*
		Base structure		*9.29*
Payload total				830.1
	Deck total			198.1
		Deck NSM		*170*

		Deck structure	*28.1*
	Deck rib structure		23.29
	Thruster support		1.75
	Tank support rods		0.42
	Lumped masses		606.5
Separation fittings			32.11
Cruise stage total			**2730**
	SEP cylinder total		144.6
		NSM	*114.82*
		Structure	*29.78*
	Thrust tube total		168.2
		NSM	*78.32*
		Structure	*89.89*
	Probe support		31.32
	Hydrazine tank support		3.4
	SEP Engine support		30.28
	Solar array support		3.54
	Solar arrays		400.4
	XE tank support		15.61
	Lumped masses		1932
Total stackup			**4190**

POST-STUDY EVALUATION WITH HYPERSIZER™

Due to the numerous iterations involved in integrating the required system/payload components and associated support structure into the available volume allowed by the 2.88m flattened ellipsled shape, structural mass optimization was not performed within the original design schedule. Shortly after the systems review for the Neptune Aerocapture study (October 28, 29, 2003), further analysis was performed on the 2.88m ellipsled using HyperSizer™ in an effort to realize some gains by optimizing the ellipsled structure mass. As discussed earlier, HyperSizer™ reads in the FEM mesh and internal loads from an outside FEA (in this case I-DEAS), then steps through a user-defined design space, applying the internal loads to local model regions called panels. While not a true optimizer, HyperSizer™ uses closed form solutions to step through all user-specified material, size, and construction method permutations for each model panel to find the lightest structure to pass all strength and stability requirements. This can result in adjacent panels having totally different sizing or construction techniques. While the result may yield the lightest possible structure, it is often not a manufacturable one. The user may then need to adjust the design space or link certain panels for the sake of manufacturability and rerun HyperSizer™. Lastly, since HyperSizer™ only checks local panel buckling modes and natural frequencies, the full FEM must be re-evaluated in the FEA code for global stability and natural frequencies.

For the mass optimization on the ellipsled structure, only the sandwich construction family of panels was looked at. This was primarily due to previous experience with this type of structure[15] and due to time limitations. As a result of the HyperSizer™ analysis, the ellipsled structure mass was reduced by 39.1 kg, from 134.4 kg (the upper thruster frame and propulsion tank supports were not evaluated) to 95.3 kg. The mass reduction was realized by sandwich thickness and/or face sheet reduction in some locations, and in blade stiffener thickness reductions. The first pass through HyperSizer™ showed a 56.2 kg mass reduction, but all of this could not be realized when adjustments were made for structure manufacturability. The resulting structure was re-evaluated in I-DEAS to check global stability and natural frequencies. As a result of reducing mass without significant stiffness reduction, the overall stackup natural frequency climbed slightly from 11.08 Hz to 11.84 Hz. For the ellipsled only at aerocapture, the global buckling margin of safety increased from 1.97 to 2.51. For the full stackup, the global buckling margin increased from 2.47 to 3.15.

CONCLUSIONS

A successful aerocapture mission at Neptune depends on success of many subsystems, including structure that will house and support the required payload, sustain launch loads, sustain aerocapture inertia loads and heating, and provide all of the above with a minimum mass. The structural analysis portion of the Neptune aerocapture systems design study showed that the chosen stackup design of a stiffened-skin construction PM/cruise stage supporting a 2.88m ellipsled aerocapture vehicle is a feasible approach when using a Delta IV Heavy launch vehicle, and that the stiffened sandwich ellipsled structure design is a feasible approach for aerocapture at Neptune. The resulting structure masses were within system allocations and allowed a total spacecraft mass that would meet the mission requirements. The results of this study may serve as a starting point for more refined analyses of a Neptune aerocapture ellipsled and cruise stage. In addition, several observations were made from the study results:

1. The flattened ellipsled shape was volumetrically inefficient in that CG requirements pushed components towards the bottom of the ellipsled, leaving the upper portion largely unused.

2. The MEL was under constant revision and was not connected to a 3D model that could be imported into I-DEAS. Analysis and MEL should be completely integrated to allow the analysts the most updated design information.

3. The aeroshell sizing and payload support structure sizing were strongly linked, and required numerous separation fittings to provide load paths from the payload deck to the aeroshell. Further analysis and optimization is warranted to help reduce this separation system complexity.

4. The use of HyperSizer™ sizing software in this study demonstrated its capabilities to the design study team and displayed how it may be applied to ellipsled geometry. The software greatly reduced analysis time by using the same finite element mesh for many trial configurations. Typical FEA modeling of bladed stiffened panels would have the analysts modeling separate stiffeners and requiring a remesh after each solution of the model. HyperSizer™ avoids this and allows many trial iterations in one solution. Further mass reduction may be possible by applying HyperSizer™ to the cruise stage structure.

ACKNOWLEDGEMENTS

The author wishes to acknowledge the following people for their contributions to the Neptune aerocapture system structural analysis: Nora Okong'o and Rob Bailey (JPL) for their work in packaging the ellipsled and cruise stage payloads within the constraints of the structural layout; Bernie Laub (ARC) for the TPS sizing on the ellipsled forebody and backshell; and Glenn Hrinda (LaRC) for the initial structural sizing on the 5.5m aeroshell and for guidance in using HyperSizer™.

REFERENCES

[1]Lockwood, M.K., "Overview", Neptune Aerocapture Systems Analysis Review, Marshall Space Flight Center, Huntsville, AL, October 28, 29, 2003.

[2]"Delta IV Payload Planners Guide", The Boeing Company, Huntington Beach, CA, 2000.

[3]Starr, B.R., and Powell, R.W., "Simulation, Monte Carlo, Performance", Neptune Aerocapture Systems Analysis Review, Marshall Space Flight Center, Huntsville, AL, October 28, 29, 2003.

[4]EDS PLM Solutions I-DEAS software versions 9 and higher, Electronic Data Systems Corporation, Plano, TX.

[5]Collier Research Corporation, HyperSizer™ Structural Sizing Software, Book 1: Tutorial & Applications, Second Edition, Collier Research Corporation, October 1998.

[6]Justus, C.G., Duvall, A., and Keller, V., "Atmosphere", Neptune Aerocapture Systems Analysis Review, Marshall Space Flight Center, Huntsville, AL, October 28, 29, 2003.

[7]Noca, M., "Mission Analysis", Neptune Aerocapture Systems Analysis Review, Marshall Space Flight Center, Huntsville, AL, October 28, 29, 2003.

[8]Edquist, K.T., "Configuration & Aerodynamics", Neptune Aerocapture Systems Analysis Review, Marshall Space Flight Center, Huntsville, AL, October 28, 29, 2003.

[9]Laub, B., and Chen, Y.K., "Preliminary TPS Sizing for Neptune Aerocapture", PowerPoint Presentation, April 10, 2003.

[10]Hollis, B.R., and Olejniczak, J., "Aeroheating Environments", Neptune Aerocapture Systems Analysis Review, Marshall Space Flight Center, Huntsville, AL, October 28, 29, 2003.

[11]Hoffman, D., and Rea, J., "Aerodynamic Stability Analysis", Neptune Aerocapture Systems Analysis Review, Marshall Space Flight Center, Huntsville, AL, October 28, 29, 2003.

[12]Laub, B., and Chen, Y.K., "Thermal Protection (TPS)", Neptune Aerocapture Systems Analysis Review, Marshall Space Flight Center, Huntsville, AL, October 28, 29, 2003.

[13]Lam, J., "Spacecraft Structure", Titan Aerocapture Systems Analysis Review, Jet Propulsion Laboratory, Pasadena, CA, August 29, 30, 2002.

[14]MSC/NASTRAN Quick Reference Guide, The MacNeal-Schwendler Corporation, 1992.

[15]Hrinda, G.A., "Structure for the 5.5 m Ellipsled", Neptune Aerocapture Systems Analysis Review, Marshall Space Flight Center, Huntsville, AL, October 28, 29, 2003.

[16]Bailey, R.W., Okong'o, N., Spilker, T., and Dyke, R.E., "Spacecraft Configuration", Neptune Aerocapture Systems Analysis Review, Marshall Space Flight Center, Huntsville, AL, October 28, 29, 2003.

www.ingramcontent.com/pod-product-compliance
Lightning Source LLC
Chambersburg PA
CBHW081725170526
45167CB00009B/3707